INSULTINGLY STUPID MOVIE PHYSICS

Hollywood's Best Mistakes, Goofs and Flat-Out Destructions of the Basic Laws of the Universe

TOM ROGERS

SOURCEBOOKS HYSTERIA™
AN IMPRINT OF SOURCEBOOKS, INC.®
NAPERVILLE, ILLINOIS

Published by Sourcebooks Hysteria, an imprint of Sourcebooks, Inc.
P.O. Box 4410, Naperville, Illinois 60567-4410
(630) 961-3900
Fax: (630) 961-2168
www.sourcebooks.com

Rogers, Tom
  Insultingly stupid movie physics : Hollywood's best mistakes, goofs, and flat-out destructions of the basic laws of the universe / by Tom Rogers.
       p. cm.
  ISBN 978-1-4022-1033-4 (trade pbk.)
  1. Motion pictures--Miscellanea. 2. Physics--Miscellanea. I. Title.
  PN1998.R569 2007
  791.43--dc22
                                 2007033901

Printed and bound in the United States of America.
                VP 10 9 8 7 6 5 4 3 2 1

To the world's physics and engineering teachers, who on a regular basis stand before an often difficult crowd and share their passion for the subject. To Dr. Thomas Thorpe, whose high school lecture opened my eyes when he demonstrated momentum by knocking the podium on the floor. And to the best teacher I ever had, Professor P. K. Stein.

CONTENTS

Acknowledgments

Few large projects get done without the support of others and this book is no exception. My son Scott was invaluable for his many insightful suggestions in the preparation of the book. He patiently read and edited my drafts repeatedly. Likewise, without the tireless support of my wife and soul mate Sandy, I could not have succeeded. Not only did she repeatedly proofread the developing manuscript, she was also a constant source of positive support. My son Mark and daughter Kelly also provided numerous suggestions, as did Bill Burns.

Introduction

Contrary to the notion that art should resist the intrusion of science, this book steadfastly maintains that the art of movie-making should embrace the science of physics, because at a gut level people understand physics better than is commonly believed. (Does any living soul not have first hand experience with force, acceleration, velocity, gravity, etc.?) What's more, with a little study and reference materials like this one, it's really not all that hard to understand basic physics at an intellectual level.

In spite of its esoteric reputation, physics books are actually fairly popular—the science sections of bookstores are full of them. People are fascinated by cutting-edge topics such as string theory, black holes, dark matter, and the weirdness of quantum physics. Oddly, it's the more straightforward classical physics that often gets ignored, even though it's extremely useful and the foundation of almost everything in modern science and technology.

Once again, this book breaks with the norm by concentrating almost exclusively on the commonplace physics principles taught in just about any good high school physics course. Even though many of these physics principles date back hundreds of years to the time of Newton (his major work, *Philosophiae Naturalis Principia Mathematica*, was published in 1687), they are still incredibly relevant and fascinating in their own right.

Yes, there are equations and calculations inside for the mathematically inclined, and as a tool of understanding for physics students taking introductory courses. Hopefully, the book will give them a deeper appreciation of the subject. For the budding filmmaker or science fiction writer, the book is filled with all kinds of useful details. However, most of the calculations and heavier technical detail are enclosed in boxes so that the casual reader can skip over them. Extra detail that does not require a mathematical background is enclosed in shaded boxes, again for easy scanning.

Those who spend some time with the book will learn to see movies more clearly, understand the world around them better, and hopefully have a lot of fun in the process.

CHAPTER 1

THE NOBLE CAUSE:
Striking a Blow for Decency in Movie Physics

IT'S ONLY A MOVIE

"It's only a movie," is often spoken by fans in defense of flicks with flaky physics, as though reviewing movies for physics content is an insult. But isn't the fact that Hollywood thinks they can feed us stupid physics the real insult? Let me explain why reviewing movies for something they need is not insulting, or unnecessary—starting with a hypothetical. Imagine a football movie: a group of plucky individualists have been forged into a team by the tough yet big-hearted coach. No one gave them a chance; yet, here they are in the big game playing their hearts out as Murphy, their beloved teammate, lies in the hospital with bandaged eyes, listening to the contest via radio.

The team is behind and desperate. It's the seventh down in the eleventh quarter, so they punt a touchdown pass from the 127th yard line. But wait, this isn't football. It's nonsense. Anyone with football knowledge would think it was ridiculous; some would be offended. The scene would never appear in a movie—not because it's unlikely or hackneyed, but because it's unthinkable to take artistic license with the rules of football. (For those who don't favor American-style football, substitute basketball, soccer, hockey, or just about any other team sport. With a few modifications, the plot will still work.)

Artistic license isn't a driver's license; it's an ambulance license. It grants the right to break rules without suffering petty penalties like traffic tickets. But rule breaking can cause errors, leading to serious penalties: wrecks. Rule breaking requires care; it's not a good idea unless there's a good reason. Hollywood would never take such a gamble with the manmade rules of football. So, when it comes to something profound like the guiding rules of the universe, why, of course, break the rules at will—no risk here.

Okay, I realize that Hollywood isn't likely to reform, but at least by discussing bad movie physics it's possible to repair some of the damage done to our clear thinking by constant exposure to foolishness. Sadly, Hollywood has a rational reason for affording more respect to the rules of football than the laws of physics: audiences are more likely to know them. Ironically, movies may be part of the cure for this ailment: Hollywood's bad physics examples are good physics teaching tools. Besides, movies are almost as entertaining as physics, so what could be more fun than combining the two?

In 1997, after years of watching one Hollywood physics wreck after another, I took a stand for decency in movie physics by founding what has become the premier movie physics site on the Internet. Since American moviegoers are used to rating systems warning of possible affronts to their sensibilities from strong language, violence, and sexuality, and since warning systems are, of course, highly effective deterrents, how could I resist? I created a similar system to warn about affronts from bad physics. Well, maybe ratings aren't so effective but at least they're fun.

Movie Physics Rating System

★ GP = Good physics in general
★ PGP = Pretty good physics (just enough flaws to be fun)
★ PGP-13 = Children under 13 might be tricked into thinking the physics were pretty good; parental guidance is suggested
★ RP = Retch
★ XP = Obviously physics from an unknown universe
★ NR = Unrated. When a movie is obviously a parody, fantasy, cartoon, or is clearly based on a comic book, it can't be rated but may still have some interesting physics worth discussing.

THE IMPACT ON ARTISTIC QUALITY

To understand when the rules (the laws of physics) should not be broken, it's best to start with the situations where they can or should be. These include cartoons, parodies, and fantasies. Even top-notch science fiction routinely stretches the boundaries of physics for the sake of story.

Time-travel is a good example of acceptable physics-bending for the sake of story. Ask ten physicists about time travel and you'll get eleven different answers, and that's with two abstaining. The truth is nobody really knows for sure if it's possible, let alone how to do it. Without it, however, there would be no Terminator movies, a definite loss of some great cinematic moments (not to mention catchy gubernatorial campaign slogans).

In *The Terminator* [PGP] (1984), computers/machines have developed consciousness and a need for entertainment along with it. What to do: work a few math problems—for a computer, how mundane—or kill off humanity? It's a no-brainer: kill people. Unfortunately, those irascible humans are unenthusiastic about extinction. A human leader steps forward and pulls together an effective resistance movement. To remedy this affront, the machines send a terminator—a metal robot covered with living tissue (Arnold Schwarzenegger)—back in time to assassinate the resistance leader's mother and snuff the movement before it starts. The humans, somehow, get wind of the plot and send back one of their own to protect the mother. Both protector and terminator arrive naked since, according to the movie, anything nonliving has to be surrounded by living tissue in order to be transported backwards through time. (Evidently hair, dead skin, and fingernails are the exception.)

Okay, the business about having to surround metal with living tissue and only send naked people back in time has no scientific basis, but it's necessary for the film's central conflict. If the human could carry a futuristic weapon, he could easily blow away the terminator and spoil the fun. Instead, he's a rabbit desperately trying to avoid the jaws of a bloodthirsty wolf in possibly the highest intensity chase ever filmed.

The nakedness also taps into the deepest levels of the human psyche. Imagine arriving naked, not just at work or school but in an entirely different era. The moviemakers do the arrival scene right: they depict a gray area of physics, time travel, with a minimum of scientific mumbo-jumbo and considerable artistic purpose.

It's another matter to defy well-established physics principles for no good reason. Any bright high school physics student (probably not a target audience) can easily spot such foolishness. Many people feel it—like an annoying itch—even when they have no physics background whatsoever. They may not be able to verbalize reasons but have experienced gravity, velocity, acceleration, force, and energy firsthand their entire lives. Individuals with military experience—shooting guns, setting off explosives, flying helicopters—are especially cranky about the itch. Here's a scary thought: in spite of physics' reputation for difficulty, it's really not all that hard to learn; verbalizing soon follows.

Guidelines for Safe Movie Physics

1) Never break laws or principles taught in high school or first year college classes unless the movie is obviously a:

★ Parody
★ Fantasy—including cartoons
★ Comic book adaptation

Keep in mind that most of these physics principles have been around for decades—some for centuries. Lots of people know about them.

2) It's okay to occasionally stretch physics knowledge beyond its current boundaries when all of the following four conditions are met:

★ The stretched area of physics is not fully understood and is at least remotely possible.
★ The story cannot be done without the stretch and the stretch creates unique entertainment or artistic opportunities.
★ The stretch is explained with a minimum of scientific mumbo jumbo.
★ The stretch does not obviously contradict the first law of thermodynamics (see Chapter 3).

People do at times watch movies for mindless entertainment, but they also watch to vicariously expand their experience. It's like gaining an additional lifespan. For about two hours they can be a criminal, a saint, a drugged-out loser, or a charismatic genius without all the messy consequences. Movies provide emotional release. Vicariously blowing up an evil emperor's death star helps release the pent-up emotions from not blowing up a discourteous driver's SUV. Oddly, even a total fantasy must seem real. Moviemakers know this and go to great lengths in such areas as set and costume design to give the illusion of reality.

The *Passion of The Christ* [NR] (2004), depicting the crucifixion of Jesus, was a box office hit at least partly for this reason. For example: fake beards were applied meticulously one hair at a time. Had moviemakers applied beards the way they apply physics, the hair would have been drawn with magic markers—phosphorescent orange ones just to make sure the whiskers were exciting.

Its not that filmmakers are neglectful or unappreciative of movie physics—never. They love it, to the point of creating movie physics' very own body of clichés: exploding cars, flashing bullets, visible red laserbeams, gasoline lit with cigarettes, and so forth. These serve the cause of movies almost as well as verbal clichés serve the cause of literature.

While they last, clichés are a filmmaker's joy but eventually get moldy and have to be tossed. *Tora! Tora! Tora!* [PGP] (1970) carefully depicted the bombing of Pearl Harbor and was awarded an Oscar® for best visual effects, along with four other nominations including one for sound. The sound track used copious quantities of canned ricochet noises—which sound fake to anyone who has heard real bullets ricochet—but then the fake sounds were a standard movie physics cliché. Today, these sounds have been updated. Movies like *Saving Private Ryan* [GP] (1998), which depicted the D-Day battle with realistic bullet-sounds, have ruined this cliché's chance for a comeback.

Since successful movies require the feeling of reality, why would any moviemaker allow indecent physics into his or her film? The tools to do it right—such as computer generated imaging (CGI) software with realistic physics algorithms—exist. But generating excitement in action scenes with realistic physics takes more time and thought because it imposes more constraints. With less freedom in special effects, moviemakers might be forced to—gasp—work harder on acting, plot, and dialogue.

THE DANGER TO CLEAR THINKING

Artistic perfection is a worthy goal, but the real reason for improving movie physics is the fact that many people actually believe it, even as they say, "It's only a movie." The foolishness works its way into our collective knowledge as fact, reinforcing major misconceptions of physics along the way. These have to be unlearned before the subject can be mastered. Okay, learning physics may not be at the top of everyone's to-do list, but it is rather helpful for designing cars, computers, refrigerators, television sets, and all the bazillions of other items modern society is based on.

Hollywood moviemakers are masters of making audiences see only what they're supposed to believe. Take the case of the reader who wrote to set me straight about, arguably, the most famous vehicle "jump" ever, in *Speed* [PGP-13] (1994).

A madman has placed a bomb aboard a city bus filled with unsuspecting riders. The bomb will demolish the bus if it slows below 50 mph (80.5 km/hr). Just when disaster seems certain, a heroic cop perilously jumps aboard and bravely takes charge. After numerous near disasters on congested streets, his superiors—typical bosses who, of course, always know best—unwittingly direct him onto an uncongested route containing an unfinished overpass bridge with a 50-foot gap. Naturally, to keep from exploding, the heroic cop orders the bus's driver to speed up and jump the gap— doing the impossible to compensate for his bosses' directive.

I had written that the jump couldn't have been made on the bridge as shown in the movie. My reader wrote that if I watched in slow motion I'd see that the jump was actually done. Sure enough, when viewed this way, there are images of a bus flying through the air—exactly what the moviemakers wanted viewers to see. But look closer: the incline or ramp required for jumping is nonexistent. At the gap, the bridge is flat. Even traveling at 70 mph (113 km/hr), when the bus reached the far side of the gap, the bus would have been about 3.8 ft (1.16 m) below the roadway, slammed into the edge, and been blown apart by the bomb. (See Chapter 8 for more detail on speed.) As for the missing section of bridge, look even more closely: it's possible to see its shadow on the ground below. There was no gap. It was created on film with thousands of dollars worth of special effects.

Even thousands of dollars worth of special effects couldn't gloss over flaws in *The Core* [XP] (2003). The entire movie was based on an absurd premise: a sinister earthquake weapon, developed by an unscrupulous scientist working for misguided military men, had stopped the rotation of the Earth's core thereby disrupting Earth's magnetic field. A group of stereotypical movie heroes, inside a magical ship, must bore down to the core and restart it. In the process, they risk their lives braving the horrifying dangers of various petty human conflicts, not to mention other inconveniences like pressures and temperatures exceeding those directly beneath the Hiroshima atomic bomb blast.

It seemed obvious *The Core* did not contain rigorous science. Yet, in response to my review, a self-proclaimed scriptwriter indignantly wrote that "the science in *The Core* was mostly accurate." His key argument: the moviemakers had retained a PhD scientist as a consultant.

In reality, the moviemakers retained not one but three respected planetary scientists (all with PhDs). They were used for the momentous task of providing background information like: ". . . . the scale and size of things, how hot it gets in the core and what kind of material could conceivably withstand such a temperature." [1] (Hint: There isn't one.) Who cares that the science consultants didn't sit in the director's chair and didn't have real authority over the movie's content. Just being there said . . . well, that they were there. The man who did sit in the director's chair (Jon Amiel) and did have authority over movie content described it as: "...a little bit of science, a certain amount of fact and a lot of fiction." Strangely, the movie's science accuracy didn't seem to top his list of concerns. He is, after all, a boss and may have bungled the description, but then the PhD science consultants were mysteriously silent about correcting him.

California Institute of Technology physicist Dr. David Stevenson was not so silent. At about the time the movie came out, he had actually published a paper in the prestigious scientific journal *Nature*—to the glee of *The Core*'s producers who smelled publicity—outlining a way to send an unmanned probe into the core. Although some of his colleagues scoffed at it, the proposal had a serious purpose: to generate discussion. However, even Dr. Stevenson admits it was close to science fiction: The Earth's crust would be split with a nuclear bomb and the crack filled with a million tons or so of molten steel topped with a small probe. The molten steel would sink towards the core, carrying the probe with it. In theory, the probe would last just long enough to send back some information before being destroyed by the extreme conditions.

The joyful producers contacted Dr. Stevenson, offering payment for public endorsements—a golden opportunity for both the movie's backers and the good doctor. He declined. It wasn't the "fantastic/ridiculous stuff about sending a manned probe to the core"; he felt this could never be mistaken for serious science because it was obvious fantasy. (Wow, if only he knew.) Instead, he primarily objected to the movie's premise that the core had stopped rotating, thereby disrupting the Earth's magnetic field— a notion he describes as "just silly." Furthermore, he had issues with the movie's simplistic explanation for how the magnetic field is generated.

Certainly, Dr. Stevenson qualified as an open minded and imaginative scientist and had a reputation as such, so his less-than-complimentary public comments about *The Core* must have really hurt. The movie's producer contacted Dr. Stevenson, yet again, and chewed him out as though he had actually taken their money. Dr. Stevenson doubts he will ever be asked to consult on another movie.

The Core's trailers alone boosted it to classic status. I was inundated with requests to review it even before it hit theaters. Rarely have I seen such excitement. *The Core* became a major contender for the distinction of "worst physics movie ever." (See Chapter 20: All-Time Stupid Movie Physics Classics) It's so bad it's good. However, asserting that *The Core*'s science is "mostly accurate" is like insisting the Earth is mostly flat. At first glance it does look that way but even a little research shows it isn't.

While not so noteworthy as *The Core*, *Pearl Harbor* [PGP-13] (2001) achieved status with its make-believe history as well as its make-believe physics (see Chapter 8). It tells the story of a young band of WWII fighter pilots stationed in Pearl Harbor just in time to survive various romantic entanglements as well as distractions, such as being strafed and bombed by Japanese planes.

Of course, the devastating U.S. defeat at Pearl Harbor isn't the stuff for a traditional Hollywood happy ending. No problem—the moviemakers merely modified history: Pearl Harbor recruits were trained to fly fighters by, no less than, the famous Jimmy Doolittle and later assigned to fly bombers in his famous raid on Japan. The actual raid did little damage in Japan but was a public relations triumph in America and a useful high note for the movie's ending.

When WWII started, Doolittle was actually in Detroit helping the auto industry prepare for wartime production. He had nothing to do with training Pearl Harbor-based fighter pilots and not a single one of them participated in his famous raid. In my movie review, I briefly commented about the moviemakers' manipulation of history. (I have had at least one veteran write me that the movie made him sick to his stomach but, as mentioned earlier, military guys tend to be cranky. Geez, I wonder why.) The physics of dropping bombs in level flight is totally different from the physics of dogfighting. So, of course, training fighter pilots for a critical bomber mission when experienced bomber crews were available made perfect sense—in dreamland. The real Doolittle raiders were all from experienced B-25 bomber crews as would be expected.

Several readers wrote, with great authority, that the movie was right. One even recommended a book on the Doolittle Raid from a reputable author who backed the movie's account. This seemed incredible and when I contacted the book's author it turned out it was. He told me his book said nothing about fighter pilots from Pearl Harbor flying in the Doolittle Raid and that the idea was preposterous. Yet, I know of at least one otherwise intelligent individual who could pass a lie detector test while describing nonexistent passages supporting the movie's false account.

Why would so many believe the movie's bogus account? It had a persuasive pseudo-explanation: Pearl Harbor pilots were the only ones with combat experience. Combine this with compelling images and dramatic dialogue overlaid with rousing music and it went straight to the subconscious as fact. Of course, such misconceptions can be jolly fun, but internalizing them doesn't exactly lead to clear thinking—but then, maybe one shouldn't set such high expectations.

Lets be honest—extensive knowledge squeezes the very joy out of misconceptions and heartlessly ruins their chances of being internalized. I have friends who can't watch a movie version of a coma patient without commenting. Their son was in a coma following a serious traffic accident and they know all too well what real comas are like. Why of course, the patient suddenly wakes up fully alert years later and within hours proceeds to use undiminished martial arts skill for revenge against bad guys just like in *Kill Bill* [NR] (2003). Yeah, right. Real comas are heart-wrenching. Their victims almost always require lengthy physical therapy and rarely completely recover.

Any form of knowledge makes it harder to digest nonsense, and physics knowledge is no exception. Blatant disregard by moviemakers for well-established physics often gives people with physics knowledge mental indigestion. But, here's a shock: physics knowledge isn't the problem—ability to discern truth doesn't lead to a life of boredom and suffering. It leads to a pleasant sensation: understanding. Internalizing cleverly packaged nonsense is the real problem; it leads to muddled thinking. Still, some may find this type of thinking, if not desirable, at least entertaining. For those who do then, please, read no further. It will spoil your fun.

Summary of Movie Physics Rating Rubrics

The following is a summary of the key points discussed in this chapter that affect a movie's physics quality. These are ranked according to the seriousness of the problem. Minuses [-] rank from 1 to 3, 3 being the worst. However, when a movie gets something right that sets it apart, it gets the equivalent of a get-out-of-jail-free card. These are ranked with pluses [+] from 1 to 3, 3 being the best.

[-] [-] [-] A major plot pretext which defies the well-established physics principles commonly taught in high school. In the best case, this brands a movie as a cinematic comic book or a fantasy.

[-] [-] Scenes which defy the well-established physics principles commonly taught in high school, and are not essential to the plot.

[-] Using an obvious stupid movie physics cliché.

[+] Avoiding lengthy mumbo-jumbo explanations when stretching physics in a gray area and using this technique only when it's required for telling the story.

CHAPTER 2

MOVIEMAKER MATHEMATICS:
How Hollywood Shoots from the Hip

COUNTING SHOTS

Figure 1: machine gun

The action hero kicks open the door and steps through with akimbo (one in each fist) .45 caliber Mac 10 submachine guns. At first the villains are dumbfounded, even though there's a dozen of them and only one action hero. In predictable ignorance they reach for their weapons. The double Macs blaze out a high calorie response at a combined rate of 2,000 rounds per minute. Bottles and mirrors explode as bullets chew up everything in sight. The action hero sweeps the room with gunfire continuously for three minutes. Eventually, when the double Macs fall silent, the bad guys look like ground beef patties.

So, where are the sidekicks with wheelbarrows? Only 1.8 seconds of continuous fire empties the thirty-round magazine of a fully loaded Mac 10. Push a button, drop an empty magazine, snap in a fresh one, and it's ready to zip off another thirty shots. Even if the action hero spends half his time reloading (of course, done off camera), in three minutes the two Mac 10s will blast out 3,000 chunks of lead weighing 15 grams each, for a total of 99 pounds (45 kg)—not including the weight of 3,000 empty cartridge cases and one hundred empty magazines now scattered on the floor. The action hero would need a sidekick with a wheelbarrow just to supply the ammo. Besides, how's the hero going to reload when he has a weapon in each hand?

The hero could reduce weight problems by using a submachine gun with a slower firing rate and smaller bullets—at a cost of appearing less macho. But then the benefits would be mostly academic. When fired, gunpowder inside a cartridge burns, producing extremely hot, high-pressure gasses, which not only propel the bullet, but also momentarily torch gun barrel internals with a white hot flame. If not allowed to cool, the moving parts in any submachine gun, Mac 10 or otherwise, will seize, similar to the way a car's engine eventually seizes when driven with an empty radiator. The short cooling time available during reloading would not prevent the disastrous temperature increase caused by three minutes of maximum rate firing. Submachine guns are lightweight devices designed for firing in a few, short, intermittent bursts.

Characters in *The Matrix Revolutions* [NR] (2003) actually did supply ammunition with wheelbarrows during the loading dock battle. Alas, the wheelbarrows weren't for realism; they were a plot device used for placing a young hero in imminent danger.

The scene depicts a group of humans who have escaped from slime-filled bathtubs controlled by machines, who now live joyfully in a bleak, underground city. They are forced to make an Alamo-like stand when the machines decide to tunnel down and wipe them out. Their Alamo is a large concrete-domed room called the loading dock, which is defended, at least partly, by enormous, humanoid robotic devices called APUs, each of which is controlled by a person strapped to its front. Two gorilla arms extend from each APU torso, with heavy-duty, automatic 30-millimeter cannons attached in place of hands. These fire pretty much continuously at an invading horde of octopus-like machines called sentinels, which conveniently stream by the thousands directly toward the cannons.

The scene's young hero—apparently underage for driving an APU—is relegated to the unglamorous task of wheelbarrowing ammunition. He waits nervously in a tunnel until the APU he is supplying runs out of ammunition. The situation is desperate. He scrambles to resupply the APU just as its driver is mortally wounded. The lad not only ends up driving the APU, but also using it to heroically save the day, at least temporarily.

It's all very thrilling, except that using wheelbarrows to supply two cannons is as effective as powering a bus with a hamster wheel. Let's assume a firing rate of six hundred rounds per minute for each APU cannon, and a cartridge similar to the 30 × 113-millimeter round used by the U.S. military in Apache helicopter gunships. Each cartridge has a mass of 447 grams. The total mass of ammunition required for one minute of sustained fire by an APU with both cannons blazing is a whopping 1,180 pounds (536.4 kg). A box containing one hundred rounds would weigh over 98 pounds (44.7 kg). With both cannons firing, this box would last a mere five seconds. Keeping the APUs supplied with ammunition would not have required wheelbarrows; it would have required a fleet of Mack trucks!

Hollywood moviemakers, traditionally, have never felt a need to count gunshots and keep them to a plausible level. In the heyday of cowboy movies during the 1950s, all handguns were six-shot revolvers, yet bursts of gunfire so regularly exceeded six shots that it became a hackneyed joke among moviegoers.

Counting shots was actually a plot gimmick in Clint Eastwood's famous portrayal of a 44-magnum toting, rule-breaking detective in *Dirty Harry* [PGP] (1971). After shooting it out with bank robbers, Eastwood approaches one lying wounded in the street. As the robber is about to reach for a nearby shotgun, Eastwood levels his hefty revolver and goes into a lengthy dissertation about whether he has fired five or six shots. He had fired six but manages to bluff the bad guy into surrendering. In a later scene, Eastwood repeats the lines while leveling his revolver at a clever psychopath who is also about to reach for a gun. Unlike the bank robber the psychopath goes for the gun, but this time Eastwood isn't bluffing. While bank robbers might be stupid, at least they're not crazy.

Today, action movies are filled with semiautomatic handguns, which can blast from half a dozen up to twenty shots before reloading, not to mention the ubiquitous, rapid-firing sub-machine guns. Audiences simply can't count shots like they used to. Nevertheless, filmmakers have actually begun responding to the jokes about firearms never running out of ammo. Movies now show guns being reloaded, at least occasionally. Eventually, Hollywood will also have to face the fact that submachine guns can't be used in sustained fire for more than a few seconds. In the meantime, there are always wheelbarrows.

THE GEOMETRY OF SHOOTING FROM THE HIP

Hollywood is as creative about the geometry of aiming shots as it is about counting them. In movies, shooters—if they're good guys—routinely hit tiny targets when firing handguns from the hip without using gun sights, while bad guys routinely miss, at least, when shooting at the hero. In reality even well-trained police officers can easily miss a human-sized target when shooting a handgun from the hip at distances over 10 feet (3 meters)—which just might explain why they're trained to use a two-handed hold along with the gun sights. A little geometry combined with some physics illustrates why.

For simplicity let's mathematically model hip-shooting as though the handgun has no recoil, the shooter's elbow is fixed in space, and his wrist has no flexibility. Bullets tend to drop due to the downward force of gravity, but at close range the drop in a handgun bullet is so small that we'll overlook it. The point of impact for the bullet, then, will be determined only by the alignment of the end of the handgun's barrel, with a line drawn between the shooter's elbow and the target. Let's call this the aiming line.

This model of shooting from the hip focuses on horizontal misses, because the critical area—the torso and head—of human-sized targets is roughly twice as tall as it is wide, so it's easier to miss in the horizontal than vertical dimension. (These assumptions may seem too simplistic, but if a simple analysis indicates that a miss is likely, then a detailed one accounting for complexities like recoil, bullet drop, hand tremors, and misalignment of the body's many joints will make missing even more likely.)

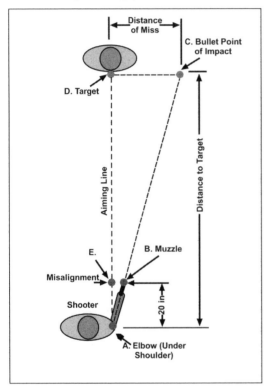

Figure 2: top view of shooter

Figure 2 shows a top view of the aiming line and the hypothetical line the bullet takes. Note the triangle formed by connecting the dots at A. elbow, B. muzzle, and E. a point directly across from the muzzle on the aiming line. Connecting the dots at A. elbow, C. bullet's point of impact, and D. target forms a similar triangle. Anyone who has taken geometry will recognize that the properties of similar triangles can be used to calculate how far the bullet misses the target, given the distance to the target, the misalignment, and the elbow-to-muzzle distance as shown in the diagram. This is calculated as follows:

(Distance of Miss) = (Distance to Target) × (Misalignment) × (Distance A to E)

If the muzzle is misaligned by only 2 inches (5 cm), the bullet will miss the target by a foot at a distance of 10 feet (3 m) from the shooter's elbow. When aimed at its center, that's enough to just graze the edge of a human-sized target. At a distance of 50 feet (15.2 m), the bullet will miss by 5 feet (1.52 m).

To understand how easy it is to be misaligned by two or more inches, try a simple experiment with a laser pointer. Hold it as though shooting from the hip, aim at a target 50 feet away, and turn it on briefly to simulate a shot. Do not move the red dot while it's turned on and note how far it misses the target. Try this multiple times from the same position on the same target.

Invariably, the first shot will completely miss. Subsequent shots will get closer until by the third or fourth, the dot may be fractions of an inch from the target. Repeated shots provide the feedback required to hit the target and may seem like a substitute for gun sights. Unfortunately, this type of feedback doesn't work as well with a handgun. First, recoil tends to misalign it, so you have to align from scratch after each shot. Second, it's often hard to see where bullets are landing. And finally, in real situations, the target is probably going to be moving or even shooting back. Shooting from the hip or without using gun sights is woefully inaccurate for most shooters, especially if the target is more than 10 feet away.

In *Pulp Fiction* [PGP] (1994) the moviemakers got just the right mix of physics precision and artistic ambiguity. During the movie two chatty hit men make a house call to remedy a misunderstanding. It seems that a misguided group has failed to uphold an agreement with the hit men's employer. After the hit men collect the overdue account, they "enlighten" the hapless souls to ensure—in the most reliable way—that there are no future misunderstandings.

Suddenly, a nervous shooter bursts from hiding across the room and empties his revolver at the startled hit men. He misses with all six shots and subsequently gets himself "enlightened," courtesy of the hit men, who do not miss when they fire back.

The scene is a classic movie moment embodying the enigmatic quality of great artwork. One hit man interprets the missed shots as a message from God wrapped in a miracle, the other as mere random chance. Had the shooter fired from the hip, random chance would have been the best explanation. But, the shooter fired in a police-style stance using both hands, making the chances of completely missing much lower. What's more, we're given a fleeting glimpse of the bullet-hole pattern in the wall behind the hit men. It looks like the bullets should have passed through the hit men, yet we can't be sure. We're left with an enigma that stirs thought long after the movie has ended.

By contrast, the two-handed submachine gun shooting, discussed earlier, is neither artful nor realistic. While extremely dangerous to innocent bystanders, it would be marginal for stopping multiple bad guys. The combination of hip shooting and jarring Mac 10 recoil would all but guarantee that the bad guys would only be hit by random chance. When the firing begins, they would scatter and dive for cover. Less than two seconds later, when the good guy held empty Macs, the surviving bad guys would pop up and serve him the ketchup.

Innocent bystanders would probably be oblivious to the danger until it was too late. A Mac 10's bullets can injure at distances up to a mile. With the exception of heavy stone, brick, and concrete construction, the bullets can penetrate walls and injure or kill people on the other side. In a big city there would be many more innocent people within the deadly range of a Mac 10 than villains.

The two-handed Mac 10–wielding action hero is the worst-case scenario for quickly running out of ammo when continuously firing a submachine gun. The best case is probably a Thompson submachine (or Tommy) gun equipped with a fifty-round drum magazine. These show up in 1930s-style gangster movies and could be fired continuously for a whopping five seconds, with an accuracy not much better than firing double Mac 10s. It's hard to believe that anyone who's serious about surviving a gun fight would shoot this way.

THE GEOMETRY OF AIMING SNIPER RIFLES

Shooting from the hip with a rifle is about as accurate as shooting from the hip with a handgun, but raise it to the shoulder, use the gun sights, squeeze the trigger—and a rifle becomes deadly accurate at much longer ranges. Carefully rest the rifle (a well-made bolt-action one with a telescopic sight) against a stable object, and it can become a sniper weapon capable of taking out targets at distances on the order of 1,000 yards (914 m).

Mathematically model rifle aiming like handgun hip-shooting, and the distance AB (see Figure 1) becomes the distance from the shoulder to the muzzle of the rifle or about 1 yard (.914 m). Misalignments with a rifle fired from a steady position and using the gun sights are likely to be less than 1/16 inch (1.6 mm) compared to the 2-inch (5 cm) misalignment likely with a handgun fired from the hip.

At 100 yards (91 meters) a 1/16-inch (1.6 mm) misalignment will cause the bullet to miss the point of aim by 6.25 inches (0.16 m). The shooter is not going to win any contests, but the bullet can still hit a human-sized target in a vital area. At 1,000 yards the same misalignment will cause the bullet to miss the point of aim by 62.5 inches (1.6 mm) or, in other words, completely miss a human-sized target.

While hitting a human-sized target at distances less than 100 yards is no big deal when firing a rifle from a steady rest position, shooting at ranges near 1,000 yards is a Zen-like mix of physical awareness, physical control, and physics designed to release incredible violence from utter stillness. Simply jerking the trigger in an amateurish way, rather than squeezing it—a skill that takes both coolness and practice—can cause misalignment on the order of 1/16 inch and a complete miss of a human-sized target. Highly skilled shooters will time the trigger pull with their muscle tremors and heart beats. Breathing while aiming is out of the question. All thoughts of the rifle butt painfully recoiling into one's shoulder, or other useless emotions, must be cast into the void. The rifle and shooter must become as one, with the single-minded purpose of sending the bullet to its target.

Sniper rifles are the Stradivarius of firearms: finely tuned, precision instruments which exactly reproduce the right vibration pattern when "played." Pull the trigger and the bullet exploding down the rifle barrel will vibrate it in a way analogous to drawing a bow across a violin string. Touch the barrel, tighten or loosen the screws holding it to the rifle's stock, and the effect is like touching, tightening, or loosening a violin string. The change in vibration can change the impact point of the bullet in unpredictable ways.

To hit the target at great distances, rifles have to be carefully sighted in and then left undisturbed until fired.

So, how does the typical Hollywood sniper practice his cold-blooded trade? He sits in front of a window, snaps open his custom brief case filled with rifle parts (each with a fitted slot in the case's foam lining), twists together the parts of his weapon with that oh-so-high-tech click, adjusts

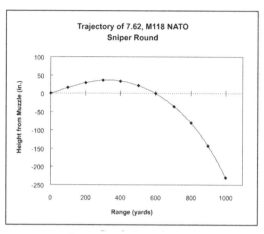

Figure 3: sniper round

the scope, aims, and fires. However, by disassembling and reassembling his instrument, he runs the risk of detuning it to a different vibration pattern, which will give the bullet a different point of impact relative to the point of aim. If he's shooting at a target less than 100 yards away (91 meters), no problem. At 1,000 yards (914 m) he'd have to spend roughly the price of a used compact car to get the engineering and craftsmanship required to create a breakdown rifle with the needed accuracy. Even then it would be questionable whether he could get it—but hey, a typical long-range assassin–(if such people exist) would probably consider it just another business expense. Besides, he'd probably do most of his work at less than 500 yards (457 m).

Even with a precision rifle and a Zen master pulling the trigger, the rifle barrel has to be elevated above the horizontal, much like the cannon on a battleship, in order to hit a target at 1,000 yards. Figure 3 shows bullet trajectories for a 7.62-millimeter NATO rifle, which is considered the world's most common sniper cartridge. When sighted in at 600 yards, but fired at a target only 300 yards away, the bullet strikes 36 inches (0.91 m) higher than the point of aim. When fired at a target 1,000 yards away, the bullet strikes 230 inches (5.8 m) low. At these distances even a gentle 10-miles-per-hour (16 kph) wind makes the bullet drift horizontally by a whopping 108 inches (2.74 m).

If the target itself is moving sideways, the sniper must lead or aim ahead of it so that the bullet arrives at the same time as the target—the longer the distance to the target, the longer the lead. If the target is moving straight away or straight toward the shooter, the task is much easier, but still not as easy as shooting a stationary target. The long-range sniper must account for all these variables, which is a daunting task.

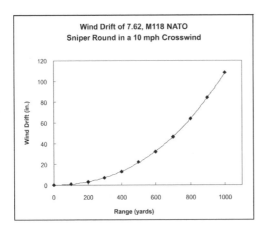

Figure 4: wind drift

On the other hand, neither bullet trajectory, wind, nor motion of the target are big problems when shooting a high powered rifle like the 7.62-millimeter NATO at ranges under 100 yards. At worst, the shot will go high by about 2 inches (assuming it's sighted in at 200 yards), or drift from the point of aim by an inch (2.5 cm) in a

10-mile-per-hour (16 kph) wind. If the target moves it will have only about a tenth of a second to get out of the bullet's path. At 100 yards the bullet has over four times as much kinetic energy as at 1,000 yards. At the closer range it is capable of blasting unimaginably gruesome wounds in whomever it strikes. A hit in the head, neck, or chest cavity will likely be fatal, even if the bullet smashes into its victim several inches from the most lethal point of impact.

According to Oliver Stone's controversial 1991 move *JFK* [RP], the Kennedy assassination was a conspiracy and could not have been carried out by an ex-marine using a scoped, high-powered, bolt-action rifle fired from a steady position, at a distance of 88 yards. Who knows? Maybe it was a conspiracy. Certainly, the CIA, mafia, communists, Cubans, police, and Girl Scouts could all be responsible. Yes, as bolt-action rifles go, Oswald's Italian-made 6.5-millimeter Mannlicher Carcano was no Stradivarius and Oswald was no sniping maestro. But while not as powerful as a 7.62 NATO rifle, the Carcano could easily have inflicted a wound gruesome enough to be fatal, nearly anywhere on a person's head. A couple inches of inaccuracy one way or the other would have been insignificant.

Oswald was no rank amateur; he was trained to shoot by the U.S. Marines. And he was shooting at a distance well under 100 yards. At that time there were thousands of hunters, shooters, and ex-military people in the U.S. who had more than enough shooting skill to carry out the heinous act. What made Oswald uniquely capable was not his shooting skill, but his pathologically cool-headed and remorseless ability to squeeze the trigger with the president's head in his gun sight.

GETTING GUNFIGHTS RIGHT

Some moviemakers do get gunfight scenes right, as in, for example, *Black Hawk Down* [GP] (2001). The movie depicts 123 elite U.S. soldiers fighting a desperate battle in Mogadishu, Somalia on Oct. 3, 1993, on a mission to capture a renegade warlord's key associates. In realistic manner the characters rarely fire anything from the hip, even when firing fully automatic weapons. Large machine guns are actually reloaded and tend to be fired in short bursts lasting no more than a few seconds at most.

One scene lends an unusual touch of realism when the hot, empty cartridge cases ejected from a rapid-firing minigun in an overhead helicopter shower down on a hapless soldier, giving him minor burns. These weapons look like old-fashioned, hand-cranked multibarreled Gatling guns, but that's as far as the comparison goes. Unlike Gatling guns, miniguns are rotated at high speed by an electric motor, which gives them an incredible firing rate. Their multiple barrels are needed to keep them from melting. Even at that, empty cartridge cases ejected from them are too hot to touch.

Moviemakers are intelligent, talented, and well funded. They can hire a busload of top experts for the price of a single supporting actor, but it does little good unless the experts are granted some power. In *Black Hawk Down* the moviemakers didn't just pay experts, they paid attention to them.

THE GEOMETRY OF DRIVING

In *The Italian Job* [PGP-13] (2003), moviemakers once again shot from the hip with respect to geometry, but in a different manner. The movie is about a happy-go-lucky group of professional criminals who steal millions of dollars worth of gold, only to be double-crossed by one of their own, who kills their leader and swipes their haul. The survivors spend the rest of the movie trying to remedy the affront.

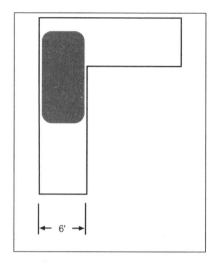

Figure 5: Mini in 6 foot wide hall

Much of the movie's plot centered on the Mini Cooper car. Supposedly, it's so small that it could be driven down the 6-foot-wide hall of the villain's house to the room where his safe was located. When the safe was opened, the Mini would be loaded with gold and driven away. This was all well and good, except that such a maneuver required 90-degree turns in the hallway and the Mini has a 17.5-foot turning radius.

While it would be jolly fun to write a mathematical proof demonstrating that the turn cannot be made, the same thing can be done with a simple scaled drawing. Figure 2 shows clearly that the turn cannot be made; there just isn't enough room.

The Italian Job actually contained a sophisticated wire-frame-style 3-D animation of the car driving through the hallway. The term wire-frame means that the visual images of the car and hallway looked like see-through objects constructed with wire-frame-like line drawings. The animation appeared to be drawn to scale and would have taken many hours to create. It could have been rotated on a computer's screen and looked at from any conceivable angle, making it easy to spot clearance problems.

There's almost no way animators could have made such a model and not known the car couldn't make the turns. At first glance it seems they would have made use of their own animation and avoided the problem by adjusting the dimensions of the hallway or eliminating the turns. Only they know why it was portrayed incorrectly. However, it's safe to say that, as is the case with many examples of bad movie physics, it was probably based on a conscious decision, rather than an accident.

Considering that physics is the most mathematical of all the sciences, it's no wonder moviemakers routinely distort it. They can't resist the urge to distort even simple mathematics.

Summary of Movie Physics Rating Rubrics

The following is a summary of the key points discussed in this chapter that affect a movie's physics quality. These are ranked according to the seriousness of the problem. Minuses [-] rank from 1 to 3, 3 being the worst. However, when a movie gets something right that sets it apart, it gets the equivalent of a get-out-of-jail-free card. These are ranked with pluses [+] from 1 to 3, 3 being the best.

[–] [–] Shooting ridiculous amounts of ammo without reloading or the barrel overheating. When really overdone, a movie starts looking like a comic book.

[–] Continuous bursts of fully automatic submachine gun fire lasting more than a few seconds.

[–] Shooting from the hip and hitting anything smaller than a man-sized target at a distance of under 10 feet (3 m).

[–] Shooting from the hip and hitting anything at a distance of over 10 feet (3 m).

[–] Sniper rifles in brief cases.

[–] Ignoring limitations imposed by simple geometry, such as maneuvering vehicles in impossibly small areas.

[+] Depicting the fact that guns get really hot from rapid fire.

CHAPTER 3

CONSERVATION OF MASS AND ENERGY:
Is Anything Sacred?

LAWS VERSUS MODELS

A few hundred years ago, the science of physics was started by an enlightened few who viewed the world as divinely created by a law-giving entity. These laws—the Ten Commandments—were provided to humanity, along with free will. Inanimate objects weren't so lucky; they just got laws. The enlightened few began to realize they could attain magic-like power by taking advantage of the inanimate world's order and lack of free will. Categorizing the universe's order into the principles and laws of physics essentially gave the enlightened few an operating manual for the universe.

In reality, the laws of physics are more like Barbie dolls than immutable laws of creation. Barbie is a model that avoids unneeded complexity and only deals with a few aspects of reality, but deals with them effectively. She is useful for modeling hair and clothing styles, in order to inexpensively predict what will look good on real people—at least on magnificently proportioned ones. Barbie lacks many real-world features such as fully workable joints, a pancreas, and a large intestine. Using her for surgery practice would be a poor application. Internal organs, however, aren't needed for understanding clothing styles.

Likewise, the laws or models of physics aren't useful for modeling everything, but are remarkably helpful in the areas where they can be applied. For example, they allow engineers to predict that an aircraft will actually fly long before it's completely built. Most modern innovations, including movie cameras, sound systems, and air-conditioned theaters, owe their existence at least in part to our understanding of the models of physics. Devices such as these have granted Hollywood the power to create entire celluloid worlds.

Not every model in physics has achieved the status of being called a law. This term is usually applied only to the most tested and reliable of models. Due to their reliability, physics' laws are never repealed, but they are sometimes refined. Often the refinement involves a better definition of the law's limitations. For example, Newton's second law does not work well when an object's speed approaches the speed of light. For that situation, Einstein's theory of relativity is needed. Yet for slower speeds, both Newton and Einstein gave the same answers.

THE FIRST LAW OF THERMODYNAMICS—A SYNOPSIS

Unlike Hollywood where nothing seems sacred, physics has a law that is as close to absolute truth as anything known to humanity. It's called the conservation of energy, and is sometimes referred to as the first law of thermodynamics.

The first law says that matter is essentially a form of energy, and that while energy can change its form, it cannot be destroyed or created. This law started out as two laws: the law of conservation of energy and the law of conservation of mass, but, when Einstein showed that mass could be converted into energy and vice versa, the two laws were combined into one. In this case, refining a law consisted of simplifying it. According to Nobel Prize–winning physicist Richard Feynman, "There is no known exception to this law—it is exact so far as we know."[2] If someone ever finds an exception it will shake science to its foundations.

EINSTEIN SIMPLIFIES THE FIRST LAW

At one time conservation of mass was considered separate from conservation of energy. Then Einstein demonstrated that mass could be converted to energy and vice versa, according to his famous equation:

$$E = mc^2$$

Where:
E = energy released when converting mass to energy, or energy
required when converting in the opposite direction
m = amount of mass converted to energy or amount of mass produced if converting in the opposite direction
c = velocity of light in a vacuum

Assume m = 1.0 kg
E = 1.0 kg (3 × 108 m/s)2
[] = 9.0 × 10^{16} J or 21.5 megatons of TNT

Fortunately, it's incredibly difficult to convert mass into energy, or even an innocuous object such as the phonebook could become a nuclear bomb. The conversion of energy into mass is likewise incredibly difficult.

Movie violations of the first law should be unforgivable, and yet they're common. In *The Hulk* [NR] (2003), a nerdy looking scientist transforms into a massive brute over two times taller, with no apparent intake of matter. The Hulk appears to grow larger as he absorbs the energy of various attempts to kill him. True, energy can be converted into matter, but the conversion is incredibly difficult and requires a massive amount of energy to produce a miniscule amount of mass. The energy output of roughly a 100-megaton nuclear bomb—the largest ever built—would create just 10 pounds (4.6 kg) of matter, assuming the conversion is 100 percent efficient. By contrast, any energy absorbed by the Hulk would be minor. Even if he could use it to increase his mass, the result would be imperceptible. For the Hulk, the only real possibility is to find a source of matter that can be easily scooped up, and air is about the only choice. Inconveniently, its density is at least a thousand times lower than the Hulk's. Absorbing enough mass to make the hulk huge and maintain his density would create a whirlwind around him, not to mention inconvenient problems with chemistry. How are a few thousand pounds of oxygen and nitrogen atoms going to be transformed into Hulk-type material?

In *Spider-Man* [NR] (2002), Peter Parker is bitten by a genetically modified spider, which imparts spider-like qualities to him. He finds he can shoot strands of spider web from his wrists. These web strands adhere instantly to objects like tall buildings, and enable Spider-Man to swing Tarzan-like while traveling great distances at fairly high speeds. Unfortunately, the web strands would also require a great deal of matter that seems to come from nowhere.

A web strand would probably need to be at least 0.5 centimeters in diameter to support Spider-Man's web-swinging antics. If such a strand were 100 meters long, it would have a volume of 0.002 cubic meters, compared to Spider-Man's estimated volume of 0.07 cubic meters. Spider-Man would lose 2.9 percent of his volume every time he shoots a 100-meter-long web. Web swinging a mere mile (1.61 km) of horizontal distance would use up about 33 percent of his body volume (assuming his web makes a 45-degree angle with the vertical at the beginning and end of each swing, and each web is 100 m long). He would be skeletal by the time he arrived and would have to eat huge volumes of food to compensate. Yet, none of this happens in the movie.

This analysis assumes that the volume of web-producing chemicals stored in Spidey equals the volume of web produced. However, even if the chemical volume were half the web volume, Spidey's volume is still going to fluctuate wildly if he does much web swinging. Yes, he could grow a spider fluid tank that could fill and drain as needed. But, assuming he continues to require human internal organs to live, where is he going to inconspicuously put the tank? On the other hand, *Spider-Man* and *The Hulk* are obviously based on comic books, so . . . okay . . . they have to be begrudgingly forgiven.

THE SECOND LAW OF THERMODYNAMICS—YET ANOTHER SYNOPSIS

The second law of thermodynamics is much more difficult to state, not to mention grasp, but is considered about as immutable as the first law. It would be easy to write a book about the second law and still not totally explain it or explore all its aspects. Frank L. Lambert, Professor Emeritus, Occidental College[3] summarizes the second law as follows:

> Energy spontaneously tends to flow only from being concentrated in one place to becoming diffused or dispersed and spread out.

Mechanical energy and electrical energy can be considered concentrated forms, while heat or thermal energy would be considered dispersed. A concentrated form of energy is like water in a container on top of a hill. Tip the container over and the water flows downward, spreading out as it goes to a lower level. For all practical purposes it would never be possible to get all of the water back in the container. Likewise, concentrated forms of energy can easily be transformed into dispersed forms, but it's difficult to do the reverse.

Heat can, figuratively speaking, be pumped uphill from its dispersed state into a concentrated form, but it can't be done with 100 percent efficiency. For example, less than 40 percent of the heat used to generate electricity in a typical coal-fired power plant actually ends up as electrical energy. The other 60 percent remains as heat and is dumped out of the power plant into the environment. In essence this is the cost for producing the electrical energy.

CARNOT EFFICIENCY—THE ULTIMATE LIMIT

Heat engines are the devices used to convert thermal energy (heat) into useful power. These include steam engines, gas turbines, and the various forms of internal combustion engines used in cars. The second law places strict limits on the maximum possible efficiency of heat engines. This maximum efficiency is called Carnot (pronounced car-no) efficiency and is calculated as follows:

$$e = (1 - T_c/T_H) \cdot 100.$$

Where:
e = efficiency in %
T_c = cold temperature at which heat is expelled into the surrounding environment
T_H = hot or elevated temperature produced within the heat engine by combustion, solar energy, geothermal energy, nuclear energy, or some other source

Actual efficiencies are a fraction of the Carnot efficiencies. The Carnot calculation does not account for real-world losses due to problems such as heat loss out the walls of the engine or any form of friction. In the case of automobiles, heat has to be removed from the walls of the engine's cylinders to keep them from welding themselves to the moving pistons inside. Yet even in a perfect world with no friction, the Carnot efficiency says that 100 percent of the energy contained in gasoline or any other fuel could never be converted into useful work by a car's engine.

Generally, lowering the temperature of the heat source rapidly lowers the efficiency of converting it to electrical energy. Current power plants, even nuclear ones, typically get efficiencies less than 40 percent. The 60 percent or more of unusable heat dumped into the environment after exiting the power plant is now at a much lower temperature.

It seems that this exiting heat could be run through yet another power plant to convert more of it to electrical energy, but the temperature is now too low. The heat ends up being wasted because the efficiency would be too low to reasonably attempt converting it to electrical energy.

When energy from a concentrated form, such as mechanical energy, is converted by friction to a dispersed form such as heat, it essentially can never be converted back. Hence, the second law says that there can never be a perpetual motion machine, except possibly in a frictionless environment. Unfortunately, friction is ubiquitous. The first law says that if a perpetual motion machine did exist, it could do no useful work on outside objects because that would drain energy out of it and eventually cause it to stop.

BATHTUBS AND BATTERIES

The Matrix [RP] (1999) was rising as a cinematic masterwork until about midway through, when it plummeted into the pit of first law violations. During the masterful first part, its main character, Neo (Keanu Reeves)—unknowingly trapped inside a vast computer simulation—begins to question his existence. He is approached by Trinity (Carrie-Anne Moss) and later Morpheus (Laurence Fishburne) who offer him the chance to find answers.

When Neo accepts, he discovers that he, along with most of humanity, actually exists in clear slime-filled bathtubs with all sorts of tubes and cables connecting him to a gigantic computerized machine system. The tubs are housed in a cavernous room tended by gargantuan mechanical tarantulas and are illuminated in part by frequent lightning-like, high-voltage discharges.

After freeing Neo and giving him lengthy rehabilitation treatments, Morpheus reveals the truth. The machines were given artificial intelligence (AI), which apparently turned them into control freaks. One might think that possessing intelligence—artificial or otherwise—would have led to understanding, but no, it led to war.

The machines were running on solar energy, so humans attempted to pull the plug by blotting out the Sun. This was very clever since humans are powered by food, which also depends on an abundant supply of solar energy. Evidently, humans stocked up on canned goods before blotting out the Sun. The machines turned the tables by enslaving humans and plugging into them as a power source.

Morpheus tells us that a human has the bioelectrical energy of a 120-volt battery; but is it a camera battery, a car battery, or something else? Volts are a measure of electrical potential energy per unit of charge, not just a measure of energy. A small 120-volt battery could provide a tiny flow of charge and, hence, a tiny amount of energy; a large 120-volt battery, a huge amount. Besides, 120-volt batteries are hard to find. Certainly, Wal-Mart doesn't carry them, so what Morpheus means when he refers to one is hard to discern.

We're also told that humans put out 25,000 British thermal units of body heat. If this happens continuously each second, it's an impressive rate of 26.4 megawatts. If the heat could magically be converted to electricity, it could power a small city. If the body heat were given off over a year, it would be a paltry rate of 0.84 watts. Even if it were magically converted to electricity, 0.84 watts would not be enough to power most light bulbs.

Unfortunately, the second law casts doubt on whether any significant part of body heat could be converted to electrical energy. Body heat is a very dispersed form of energy, while electrical energy is a very concentrated form. Body temperature is so low that converting the body's heat to electrical energy would have a miniscule efficiency. How human body heat would be useful to a vast electronic computer system is a mystery. Generally, electronics have to be cooled.

Morpheus concludes his energy discussion by lofting a copper-topped D-cell flashlight battery (ironically rated at 1.5 volts), implying that this represents the puny power output of a human. It's meant as a highly dramatic gesture, but the numbers make it look like a parody.

A bed-ridden, six-foot-tall, 160-pound, twenty-five-year-old male requires about 2,000 kilocalories worth of food energy per day just to stay alive. Note, that one food calorie equals 1,000 calories. In other words the calories reported for foods are really kilocalories. (Why they're not called kilocalories instead of capitalizing the "c" is anyone's guess.) This works out to a power rate of 96.6 watts, or about as much as a typical incandescent light bulb.

In a day's time a tub-bound human uses 2.3 kilowatt-hours of energy to stay alive. A copper-topped D-cell flashlight battery holds about 0.023 kilowatt-hours of energy. In other words, it would take about one hundred D-cells worth of food energy every day to keep a human going.

The first law clearly says that humans cannot produce more energy than they consume. Hence, humans cannot be considered an energy source. At best, they are devices that can convert food energy (a type of chemical energy) into electrical energy. If they produce the output of a D-cell, they have a best-case food conversion efficiency of less than 1 percent. However, the energy required to collect and distribute the food as well as maintain the slime tubs would be more than the human electrical output. Why would the machines bother to keep them?

Feeding liquefied dead humans (as done in the movie) back to the living ones doesn't help. Meeting human energy needs with this system would make it a giant perpetual motion machine—clearly impossible according to the second law of thermodynamics.

A 160-pound human probably contains about as many food calories as (please forgive the comparison) 160 pounds of hamburger meat. At about 1,200 kilocalories per pound this works out to 19,200 kilocalories of possible food energy. Just to stay alive for fifty years this human would have to consume over 36 million kilocalories—equivalent to ingesting 190 recycled humans, or about 3.8 dead humans a year. Where are all these people supposed to come from?

Matrix apologists have proposed that humans are not a primary power source but a backup source like the battery in a car. Here's a thought: why don't the machines just use car batteries? Had the machines been thinking, they would have raided their local Wal-Mart for automotive batteries before starting the war.

Surely the machines have some nonhuman form of energy storage in their hordes of sentinels—the octopus-like robots that float around in subterranean tunnels seeking to kill humans who've escaped their bathtubs. Sentinels would have to carry a large amount of stored energy to keep going.

To cover itself, the movie throws in a quick mention that the human energy source powering the machines is combined with a source of fusion. This is like getting on a jet airliner and having the captain explain in great detail that the plane is rubber band powered, then adding that it also has four jet engines. Guess which power source gets it off the ground? Duh.

MYTHICAL ROBOTS

A.I.: Artificial Intelligence [XP] (2001) couldn't even make it past the opening without slamming into the first and second laws. The movie opens with a scene of churning surf. The narrator proclaims that greenhouse gasses have warmed Earth, causing the ice caps to melt and flood major cities in coastal areas. As a result, populations have been displaced and "hundreds of millions" in poor nations have starved. So far it's science fiction, but not for long.

The narrator continues by announcing that prosperous nations have sustained their prosperity to a large extent by creating the perfect low-cost labor force: robots. According to the narrator, these robots require no resources beyond those used to create them. In other words, we're asked to believe that the robots never need to be recharged, refueled, or rebuilt. They are essentially perpetual motion machines, which break the first and second laws.

As mentioned in the first chapter, sometimes there are good artistic reasons to defy a law of physics. Great artists have often been defiant. Edouard Manet and René Magritte are both famous for creating paintings that look realistic but use impossible physics.

Manet defied physics to provoke the French Academy. His painting *Le Bar aux Folies-Bergère* (*The Bar at the Folies-Bergère*) deliberately shows an impossible reflection of a young woman in a mirror. The viewer is standing directly in front of the young lady. Her reflection in the mirror on the wall

Figure 6: *Le Bar aux Folies-Bergère* | *The Bar at the Folies-Bergère* | Edouard Manet

behind her should be directly behind her and almost impossible to see. Instead, Manet painted it to the far right side (see Figure 6). This no doubt horrified official art critics of the time, much to the delight of Manet.

Magritte broke the laws of physics as a type of visual joke or riddle. His painting *L'Empire des Lumieres* (*The Domain of Lights*) shows a night scene occurring during the day. The sky is noticeably in daylight while the house below is obviously illuminated as it would be at night.

There's a big difference between insightful or clever rule breaking and the clumsiness of an amateur who can't get perspective, proportions, and the overall physics of vision right. Unfortunately, the statements concerning robots in *A.I.* don't seem to be particularly clever or insightful.

The tendency to view machines as superior to their biological counterparts has been widespread. Biomedical engineering literature of the early 1970s proclaimed that science would produce a viable mechanical heart replacement for humans in about twenty years. By now, individuals with artificial tickers should be commonplace—a heart was, after all, merely a pump. How hard could it be to replace it? But even today, replacing a human heart with a mechanical device is still in the experimental phase. Yes, considerable progress has been made, but the truth is, biologically produced hearts are still vastly superior to mechanical ones.

Apologists for *A.I.* often say that the narrator in the opening scenes is not to be taken literally. What he really means is that the robots can go for extremely long periods of time without being recharged, refueled, or rebuilt. These lengthy time spans make it seem as though robots need no resources beyond those used to create them.

Ten years between refueling is about the shortest time that would give the illusion of robots needing no outside resources. So let's see how much energy storage would be needed. Gasoline is used for energy storage in automobiles not just because it's available but because it has one of the best energy to mass ratios of any energy storage medium currently available. It seems like a good starting point. As mentioned earlier, a typical male human requires about 2,000 kilocalories of food energy per day just to stay alive; if he performs manual labor, he needs about 2,500 kilocalories. Assume the human is replaced with a 100 percent efficient gasoline-powered robot. A ten-year fuel supply would require 1,900 pounds (864 kg) of gasoline. Obviously, no human-sized robot is going to be walking around with that much fuel aboard, let alone be 100 percent efficient. To be viable as a ten-year energy supply, a storage medium would have to contain about one hundred times more energy per pound than gasoline—fat chance.

The only solution would be to use some type of nuclear fuel. Needless to say, this has all kinds of problems: radioactivity, for one. Certainly, the movie's scene in which obsolete robots are destroyed in sick ways to entertain a cheering circus-like throng could never be done. The crowd would be irradiated from leftover fuel and nuclear waste inside the robots.

Nuclear fuel produces heat, which then has to be converted to electrical energy in order to be useful inside a robot. There's no known mechanism for producing electricity directly from nuclear fuel. The second law says that only a fraction of the heat produced by the nuclear fuel will actually end up as electrical energy. The rest has to be dumped into the environment as waste heat. To get any appreciable amount of efficiency, the heat converted to energy has to be produced at an elevated temperature, probably at least 932 degrees Fahrenheit (500°C). Aside from being radiation hazards, the robots would likely be fire hazards.

The movie's main character—a robot that looks like a little boy—becomes depressed because his adopted human mother doesn't love him. At the end of his emotional rope, he jumps into salt water in a suicide attempt. He's rescued, but before he can dry off he ends up piloting a helicopter-turned-submarine to the bottom of the ocean where he accidentally becomes trapped. When rescued—a mere 2,000 years later—he functions like he'd been on the bottom a few minutes. Not even his t-shirt has deteriorated. Park a car for two years, turn the ignition key, and you'll be lucky if it sputters.

Obviously, the moviemakers took the opening narrator pretty seriously. A 2,000-year supply of nuclear fuel for the hypothetical 160-pound bedridden human described earlier would add up to the energy equivalent of 1.4 kilotons of TNT. This is like a low-yield tactical nuclear device that could take out a neighborhood. Even if the child robot had half as much, combine his stored energy with 2,000 years of neglect and he'd be a walking Chernobyl waiting for meltdown.

A.I. seems to be a movie that wants to make serious statements about the nature of love and the mother-child relationship, yet, from a physics standpoint, achieves pure silliness. Its violations of the first and second laws are not consistent with its serious tone and are unforgivable.

It's easy to obey the first and second laws, especially in movies with serious science fiction themes. If it's not done, the movie becomes, at best, a science fantasy or cinematic comic book. At worst, the movie turns into pure nonsense.

Summary of Movie Physics Rating Rubrics

The following is a summary of the key points discussed in this chapter that affect a movie's physics quality rating. These are ranked according to the seriousness of the problem. Minuses [–] rank from 1 to 3, 3 being the worst. However, when a movie gets something right that sets it apart, it gets the equivalent of a get-out-of-jail-free card. These are ranked with pluses [+] from 1 to 3, 3 being the best.

[–] [–] [–] Serious violations of the first law. Includes the metamorphosis of creatures into a very large size with no apparent source of matter or reduction in density.

[–] [–] [–] Depictions of perpetual motion machines.

[–] [–] [–] Depictions of super-human robots that never need recharging, refueling, or rebuilding. At best this categorizes a film as a cinematic comic book.

[–] [–] Minor violations of the first or second laws that are not significant parts of the story line or plot.

[+] [+] Depicting robots that break down and regularly need to be refueled or recharged.

CHAPTER 4

Scaling Problems:
Big Bugs and Little People

It's an old movie gimmick: radioactive contamination, toxic waste, genetic engineering, or some other out-of-control technology abnormally shrinks or expands someone or some creature. While the gimmick is certainly entertaining, the physics are flaky.

SCALING DOWN HUMANS

In *Honey I Shrunk the Kids* [NR] (1989), Rick Moranis plays a whacky inventor named Wayne Szalinski who works at home and successfully builds a device for miniaturizing objects. Naturally, his son, daughter, and two neighbor kids end up accidentally (surprise, surprise) getting shrunk. After Szalinski unknowingly sweeps them up and throws them out, they must journey across the foreboding backyard to get back to the house. Who would have guessed that hanging around the backyard could be such an adventure?

Ordinary matter is almost entirely filled with empty space. The vast majority of mass in an atom is contained in its nucleus. The mass of its electrons is inconsequential by comparison. On an atomic level, there are huge distances between the nucleus of an atom and the nuclei of its nearest neighbors, even in a solid. Other than containing a few small specks, namely electrons, the space between nuclei is filled with essentially nothing. It is not filled with air, it's filled with void.

It's conceivable that people could be shrunk by somehow removing some of the empty space inside them. This happens in a black hole, although to a much more dramatic extent. Of course, there are many problems associated with squeezing empty space out of matter. For example, the repulsion forces between nearby atoms would increase to levels much higher than normal, making it extremely hard to keep the atoms from moving apart— explosively. However, let's ignore the problems and assume the space can be magically removed. Using this system, a 100-pound teenager reduced to the size of an ant would still have exactly the same number of atoms in her body. This means she would have exactly the same amount of mass and weight.

When something is scaled up, all its dimensions are multiplied by the scaling factor. When scaled down, all its dimensions are divided by the scaling factor. Note that any area on any object will scale up and down with the square of the scaling factor. For example, scale a human up by a factor of 10, and the area under her feet will increase by a factor of 10^2, or 100. Scale her down by a factor of 10, and the area under her feet will scale down by a factor of 100 or become 1/100 its original size.

Let's assume Szalinski's teenage daughter's foot is about 10 inches (25.4 cm) long by 4 inches (10.2) wide, giving an area of 40 square inches. When walking or running, her weight would momentarily be applied to the area of a single foot. This yields a pressure as follows:

Pressure = weight/area
= 100 lbs/40 sq inches
= 2.5 psi or 0.18 atm

If the teenager is scaled down by a factor of 100, her foot will now be 0.1 inches long by 0.04 inches wide for an area of 0.004 square inches. This increases the pressure beneath her feet by a factor of 10,000. If she places her weight on one foot, the pressure would equal 25,000 pounds per square inch. Standing on both feet, the pressure would be 12,500 pounds per square inch. Either of these pressures would easily exceed the compressive strength of concrete (typically 3,000 to 4,000 psi). But walking on a concrete surface might break her feet before it broke the concrete.

The miniature teenager will have feet with areas similar to the ends of small screwdriver blades. Place two of these vertically with the tips touching the soft soil of a typical back yard and try standing on them. They will immediately sink into the ground. Without doing any further analysis, it's possible to say that Szalinski's two kids and their two friends are never going to make it across the yard.

There's only one other conceivable way to shrink children: remove some of their molecules. Removing electrons isn't helpful because they don't have enough mass—not to mention that it would create ions and alter chemical bonds. Removing protons alters atoms into completely new materials. Pull a proton out of an oxygen atom and it becomes nitrogen. Doing this to the oxygen atoms in a room would suffocate its occupants. Removing neutrons eventually results in unstable radioactive isotopes. Take a neutron out of normal oxygen and it becomes a radioactive isotope that decays rapidly into an isotope of nitrogen. This process has a half-life of only 122.2 seconds. In other words, half of the radioactive oxygen is gone in a little over two minutes.

Removing an atom from a molecule creates a totally different compound. Taking an oxygen atom from a water molecule converts it to hydrogen gas. So removing atoms also can't be done. The only possibility left is removing whole molecules.

The question becomes, how many molecules can be removed before problems arise? Sweating, for example, removes molecules, but at some point it results in dehydration. To analyze this possibility, let's assume that the miniature person has to end up with the same density as he started with. We have to first do a little magic (called algebra) to understand this:

STARTING WITH THE DEFINITION OF DENSITY

Density = mass/volume
We can derive the following:
Mass = (volume) × (density)

This equation says that mass will be directly proportional to volume since density will be constant. For example, if volume is decreased by a factor of 10, mass will decrease by a factor of 10. Obviously, we need to know how volume scales up and down to find out how much the mass has to decrease when shrinking a teenager.

To understand this, imagine a sphere. Its volume is equal to 4/3 pi times the quantity of the radius cubed. In this case, if we scale the radius up by a factor of 10, the volume increases by 10^3, or 1000. Likewise, if we scale down by a factor of 10, the volume decreases by a factor of 10^3, or 1000. In other words it will be 1/1000 of its original size. It turns out that regardless of an object's shape, its volume will scale up and down with the cube of the scaling factor, in other words with (scaling factor)3.

If a teenager is shrunk by a factor of 100, her volume will decrease by a factor of 100^3, or 1,000,000. This means for every molecule left in the tiny version of the teenager, 999,999 molecules will have to be removed. We could, perhaps, remove cells instead of molecules, but imagine what would happen to a person if 99.9999 percent of his or her brain cells were removed.

According to the first law of thermodynamics (see Chapter 3), all the removed molecules can't simply disappear. They have to go someplace, for example, into a barrel of goo. The barrel of goo is certainly a problem, but it's only the beginning. It's going to be a nightmare figuring out which molecules in what ratios have to stay in order to end up with a working human after miniaturization. The situation gets worse when considering that humans are warm-blooded.

Like all warm-blooded creatures, humans have to eat a great deal of food just to maintain their body temperature. This food intake is proportional to surface area and scales down by the square of the scaling factor. Hence, the tiny teen's food requirement will be 1/10,000 times what it was in her full-sized form. So, what's the big deal?

Assume the normal teen eats 1 percent of her body weight in a day to maintain her temperature. Her food intake has certainly decreased, but her volume has decreased even faster since it scales down by the cube of the scaling factor. Because density was held constant, her mass and weight have decreased by the same factor as her volume. So, she will now have to eat 100 percent of her body weight every day just to maintain her temperature. Even if the ambient temperature is 70 degrees Fahrenheit (21°C), she would have to eat constantly to avoid hypothermia. This is why small critters such as ants are typically cold-blooded. It is also why tiny warm-blooded animals such as shrews have to eat constantly.

Even allowing for some magic, physics says that scaling a human down by large factors simply can't be done if it's desirable to keep her alive. Obviously, scaling problems are major flaws in movie physics and should immediately kill the chances of getting anything better than an RP rating. So why is *Honey I Shrunk the Kids* not even rated? Simple, it's a Disney family movie. There is not a speck of serious science fiction in it. It's supposed to be silly, and it is.

SCALING UP HUMANS

In a sequel called *Honey I Blew up the Kid* [NR] (1992), Rick Moranis is again creating accidental science experiments. This time, his two-year-old son Adam gets blown up to gargantuan size. The youngster is no Godzilla, but he does treat his neighborhood like a giant toy box.

Once again, there are only two conceivable ways to expand a kid. The first is to increase the space between and inside of molecules. This would likely change him into a random cloud of gas or ions that would disperse at the first sign of wind. However, let's assume that by magic his molecules stay together but with more space between them.

By How Much Could a Human Be Scaled?

To fully answer this question we'd have to define what it is to be human. A very large or very small human would likely not have the same thinking ability (it could be better or worse), lifespan, movement, or biological design as a human on the opposite end of the spectrum.

The shortest recorded man was Gul Mohammed at 1.88 feet (0.572 m) and the tallest Robert Wadlow at 8.93 feet (2.72 m), for roughly a scaling factor of 4.8 between the two.

Allowing a little extra room, it seems likely that the largest humans could be about five times bigger than the smallest ones and still live. In other words, adult human heights could range from about 1.8 feet (0.549 m) to about 9 feet (2.74 m). The large humans would have 1.5 times the differ-

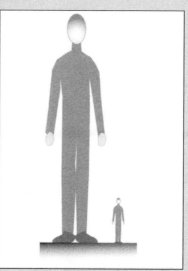

Figure 7: scaling factor

ence in blood pressure between their feet and head and likely have more health problems, including early heart failure, than those who are a more normal size of 6 feet (1.8 m) tall. (Wadlow died of an infected foot blister at age 22.) The pressure load on their joints would be 1.5 times higher than the 6-foot person, possibly leading to early arthritis. While their food intake would be less per unit of body weight, such large people would have to eat about 2.3 times as much food and spend more time finding and consuming it than the 6-foot person. Their brains would be larger than normal, but with longer nerves, bigger organs to control, more challenges to their immune systems, higher food requirements, and so forth, they'd likely not be any better at reasoning and problem solving than their normal-sized counterpart.

As for the miniature humans, while they might have some arthritis advantages, their blood vessels would be far smaller than normal and, hence, more prone to clogging. They would have to eat a greater proportion of their body weight in food to maintain their temperature and so would have to spend more time eating. With their much smaller brains they'd certainly have no advantage in thinking ability. Normal-sized humans really are near the optimum size.

In the movie, the youngster is scaled up by roughly a factor of 50. His volume increases by a factor of 125,000. Since his mass remains constant, his density will also decrease by a factor of 125,000. In other words, he will be 1/125,000 times as dense. Humans have a density similar to water (1.0 g/cc) because they are mostly water. Hence, the expanded youngster will end up with a density of 8×10^{-6} grams per cubic centimeter. This is less than 1/10 the density of hydrogen gas, which is about 9×10^{-5} grams per cubic centimeter. In other words, the title of the movie should be *Honey I Blew up the Kid and He Blew Away*.

The other alternative is the constant density approach. Add 124,999 molecules to the kid for every one he now contains. He'd have to be pumped up with an Olympic-sized swimming pool full of goo, all the while hoping the molecules would all somehow find their proper place. If we assume the additional molecules wouldn't kill him, there would be other serious problems.

We can model the youngster's leg bones as though they are vertical cylinders similar to the columns that held up ancient Greek temples. Of course, this assumes he is standing. In this position, the leg bones or columns' ability to hold up weight will be proportional to their cross-sectional area—in other words, the area of the circle that would just fit around the outside of the column.

Note that the load on a leg of a standing child, or adult for that matter, is caused by gravity. This acts in the vertical direction and is called a compression load since it tends to compress the leg. The cross-sectional area is in a horizontal plane. In other words, the cross-sectional area that determines the strength of the leg is perpendicular to the compression load's direction.

Since cross-sectional area is indeed an area, it will scale up and down with the square of the scaling factor. Hence, if the bone's diameter is doubled, its cross-sectional area and subsequent ability to support weight will increase by a factor of 2^2, or 4 (assuming no change in material properties). Obviously, its strength will increase or decrease very rapidly when scaled up or down.

When two-year-old Adam is scaled up by a factor of 50, the weight his leg bones can support will increase by a factor of 50^2, or 2,500. Unfortunately, his weight will also increase by a factor of 50^3

or 125,000. So his leg bones would have to be fifty times stronger. His neck and spine support compression loads and would also need to be fifty times stronger. But they contain numerous joints, which tend to be weaker than bones. Most likely, joints and bones all over his body would fail, and he would collapse into a very un-Disney-like mess.

King Kong

King Kong [PGP-13] (2005) features a giant gorilla that ends up rampaging around New York City. It's one of the most enduring Hollywood movie images, but how realistic is its physics?

The question can be answered in part by calculating the scaling factor. A real gorilla is, at most, about 5 feet 7 inches high (1.75 m) and 396 pounds (180 kg) in weight. Increase its height to a King Kong–sized 25 feet (7.6 m) and the scaling factor is 4.5—a value within the design range of many types of animals, such as dogs (a Great Dane, for example, is at least four times taller than a Chihuahua), or for that matter humans. Carbon-based biological creatures in the size range of King Kong—namely dinosaurs—have existed. The great ape would have a weight of 18 tons (16,000 km) compared to a brachiosaurus weight of 84.8 tons (77,100 kg) or a tyrannosaurus weight of 7 tons (6,400 kg), again within the values for real (albeit extinct) animals. Such an enormous gorilla is conceivable, but could it behave as depicted in the movie?

The muscle strength to weight ratio would be less than one-fourth its original value (assuming muscle strength scaled up with cross-sectional area.) Hence, the great beast would not have the leaping and climbing ability depicted in the movie. Many of the beast's features, such as eye and nose size, would be way over designed, representing a poor level of optimization. A giant squid is about the only animal with such large eyes, and these are designed to be able to gather enough light for vision even in the near-lightless depths of the ocean. When standing, Kong would have about 410 times more gravitational potential energy stored in his body than a gorilla. However, his bones would only be about 20 times stronger (in tension or compression). If King Kong fell during a fight, he would be prone to injuries like broken bones. Could he successfully hold an uninjured blonde in one hand while fighting three tyrannosauruses simultaneously? Yeah right. These "terrible lizards" with massive jaws full of 12-inch- (30-cm-) long serrated teeth designed to tear flesh and penetrate bone, would be fully optimized oversized meat grinders. King Kong would end up looking like ground round and probably bleed to death even if he won. As for the blonde, she'd probably be shaken and bounced to death.

ANTS, ELEPHANTS, AND ELASTIC STABILITY

The analysis of scaling humans up and down has ignored something called elastic stability. A typical column fails in compression when the weight it supports is too high. Its maximum weight is directly proportional to its cross-sectional area but is not at all influenced by column length. However, make a column too long, and length becomes an issue. The longer the column, the more likely it is to fail because it lacks elastic stability. When it fails, it will suddenly buckle and collapse with much smaller loads than those needed to cause a compression failure in a shorter column. The bones in a person's legs aren't long enough to make elastic stability an issue, but obviously it imposes a limit on how long a creature's legs can ultimately be.

Ants are famous for their ability to carry fifty times their weight. If scaled up from 3 millimeters to 3 meters Hollywood-style (a scaling factor of 1,000), they would indeed be formidable and terrifying. Of course, scaling up Hollywood-style means they could still carry fifty times their weight. To understand why this is nonsense, let's make the same analysis as was just done for the two-year-old.

The ant's leg strength in compression would increase by a factor of $1,000^2$, or 1 million. However, the ant's weight would increase by a factor of $1,000^3$, or 1 billion. The ratio of weight to leg strength in compression would increase by a factor of 1,000. Each leg in the ant's large-size version would have to be 1,000 times stronger than needed in the ant's original size. Most likely, the ant would collapse under its own weight. However, the scenario gets even worse. Ant legs are very long and thin. If scaled up, they would have elastic stability problems as described earlier.

The effects of elastic stability can be demonstrated with a soda straw. Cut a 2-centimeter-long piece. Compress it like a leg bone by squeezing it between your thumb and index finger. It takes a lot of force to make it fail. Conduct the same experiment with a 12-centimeter-long straw and you'll find that the longer straw buckles immediately with the slightest side force or misalignment. The longer straw has elastic instability.

The long, thin legs of ants, spiders, and other crawly little critters would be a complete disaster in an oversized version. The simple truth is that the design of creatures is dictated in large part by the physics of their size. They just can't be scaled up or down by much without a design change. That's why elephants have proportionally larger legs than goats.

Most of the incredible strength of critters such as ants and spiders is related to their small size. These amazing abilities simply do not scale up. If a magical spider bit a human and turned him into a man-sized spider-person with properly scaled-up spider strength, he would be lucky if he could get out of bed.

SCALING MACHINERY

In general, things get proportionately weaker from a mechanical standpoint and considerably heavier when scaled up. A multiple-engine bomber will have proportionately larger wings than a fighter plane in order to provide the additional lift required to counteract the bomber's much higher weight. Due to its smaller size, the fighter will be able to make maneuvers that would tear the wings off a bomber.

This table shows a comparison of a P-51 single-seat fighter aircraft to a four-engine B-17 bomber, both WWII aircraft. Actual scale-up factors are the bomber's values—wing span, length, height—divided by the fighter's values. In a Hollywood-style scale-up, every dimension on the aircraft would be multiplied by a constant scale-up factor.

Table 1: Comparison of WWII Fighter to Bomber

	P-51	B-17G	Actual Scale Up Factors	Hollywood Style Scale Up
Wing Span	37.4	103	2.8	2.8
Length (ft)	32.2	79.8	2.5	2.8
Height (ft)	13.7	19.1	1.4	2.8
Weight (lbs)				
Empty	7125	35,800	5.0	2.8^3 = 22,952
Operational	11,600	65,600	5.7	2.8^3 = 22,952
Max Speed (mph)	437	323	0.74	2.8
Crew	1	10	10	-
Total Power (hp)	1590	4800	3.0	-

Figure 8: fighter to bomber comparison ratio

As a result, parameters such as weight would scale up by the cube of the scale-up factor. Scaling up motor horsepower is more complex. If it scaled up with the engines displacement volume, motor horsepower would be increased by a factor of 22,952. Generally, Hollywood assumes that performance factors such as speed would also increase by the scale-up factor, following the principle that bigger has to be better. In this case, such an assumption would be especially silly since it would make the bomber supersonic, and supersonic aerodynamics are completely different from subsonic.

Note that actual scale-up factors are inconsistent with the Hollywood factors. In other words, a WWII bomber is not simply a scaled-up fighter. It has to be redesigned specifically for its mission. Some aspects of performance, such as maneuverability and air speed, have to be sacrificed to gain others, such as payload.

Likewise, the pilot of a bomber is not going to be interchangeable with the pilot of a fighter, and vice versa. While both men may be capable of flying the other's craft, it's going to take a lot of hours of flying to master it. The business of using fighter pilots to fly a bomber mission as depicted in *Pearl Harbor* (see Chapter 1) is a Hollywood fantasy.

This chapter has dealt with only a handful of scale-up issues. The problems extend into all areas of science and engineering. For example, physicists currently use quantum mechanics on the atomic level and general relativity on an astronomical level; the two are not interchangeable. Efforts have been repeatedly made to combine them into a set of universal principles. Eventually, this may be done. Meanwhile, even the principles of physics are subject to scaling issues.

Summary of Movie Physics Rating Rubrics

The following is a summary of the key points discussed in this chapter that affect a movie's physics quality rating. These are ranked according to the seriousness of the problem. Minuses [–] rank from 1 to 3, 3 being the worst. However, when a movie gets something right that sets it apart, it gets the equivalent of a get-out-of-jail-free card. These are ranked with pluses [+] from 1 to 3, 3 being the best.

[–] [–] [–] Scaling a living creature up or down by incredible amounts without killing it. This almost always tags a movie as comic-book-like when done as a major plot device.

[–] [–] Claiming that the incredible strength of small critters, such as spiders, can be passed on to a much larger critter, such as a human. Again, this tends to tag a movie as comic-book-like. Of course, if it is based on a comic book, then it's not necessarily a problem.

[–] Portraying significantly larger or smaller inanimate objects as though they will behave similarly to their regular size.

CHAPTER 5

Inertia and Newton's First Law:
Why Blowing Up Spacecraft Is a Bad Idea

NEWTON'S FIRST LAW—A BRIEF HISTORY

Newton's first law could be called the bunny principle—it states that objects keep going and going, just like the Energizer bunny, until a net force changes their motion. The change could be speeding up, slowing down, or a change in direction.

Newton's first law is more typically called the law of inertia, since inertia is the property of mass that resists a change in motion. The higher the amount of mass, the more difficult it is to get an object moving. Once moving, the higher the mass, the harder it is to stop or change the object's direction.

It took geniuses to figure out this simple principle; Leonardo da Vinci was one of the first to do so. Human experience tells us that a force is required to keep an object moving. When the ox quits pulling, the cart stops; so it seems obvious that a force is needed for motion. But, what if there were always a force present that took over and caused stopping as soon as the pulling force ended? In our world, it's essentially impossible to be totally free of friction or some other resistance force, such as air resistance. So, objects do tend to stop if nothing is pushing or pulling them in the direction of motion, but it's because there's virtually always a resistance force pushing in the opposite direction of the motion.

The difference between the resistance force and pushing force is called the net force. If the pushing force is larger than the resistance force, the object speeds up. If it's the opposite, the object slows down. If the two forces are exactly the same size, the object stays at constant velocity. If it's moving, it continues to move. If it's stationary, it remains stationary.

Da Vinci most likely understood the ubiquitous nature of resistance forces like friction. He couldn't create a frictionless environment, but could imagine one. He might have visualized a friction-free world with only a single moving object in it. In other words, there would be no other objects for it to collide with and no resistance forces to alter its motion. He would then have asked a key question: what happens to the object's motion over time? The answer: nothing. The object would move in a straight line at the same speed forever. This profound principle was called Leonardo's law.

Then along came Newton. He embellished Leonardo's law with two additional related laws, invented calculus, and managed to get Leonardo's law named after himself. It's now known as Newton's first law.

WWII NAVAL BATTLES AND THE PROTECTIVE NEWTONIAN SHIELDS

Both the concept that it takes a net force to change an object's motion and the idea that there's almost always some sort of resistance force opposing motion are profound. Let's take a look at an example to see why.

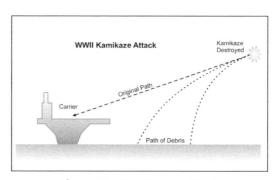

Figure 9: Kamikaze attack

Imagine a WWII battle, and a massive convoy is moving slowly through enemy waters with gun crews on alert. There are destroyers, cruisers, battleships, and aircraft carriers seemingly in every direction. Suddenly a lone aircraft swoops from the sky, diving straight for the deck of an aircraft carrier. Every gun crew on every ship that sees the aircraft opens fire. At the last moment the aircraft comes apart in a ball of fire and falls harmlessly into the sea. The aircraft carrier is safe.

Careful observation would show that the aircraft continued its forward motion even as it broke into pieces. The forces exerted by the bullets, even cannon shells, are not high enough to stop the aircraft's forward motion. They do, however, punch it full of large holes that damage the airframe, disable the control systems, wipe out the engine, and rupture the gas tanks.

With the engine disabled and the aircraft falling apart, air resistance slows the pieces and gravity pulls them downward as though the ship were protected by a giant Newtonian shield. As long as the aircraft comes apart well before it reaches the aircraft carrier, the shield works and the pieces fall harmlessly into the ocean. But the shield has a weakness—if the aircraft gets too close before coming apart, it hits the ship and does a lot of damage.

Holding together just long enough to hit a ship was the principle behind kamikaze suicide attacks. Even if the airplanes were shot up, if they got close enough, they could still do damage. In WWII, antiaircraft guns often could not shoot a plane down at a safe distance, and so kamikaze attacks did indeed inflict lots of damage. Of course, the damage was not just caused by the hurtling aircraft's mass but also by the explosives it carried.

If a kamikaze plane, or for that matter a bomb or torpedo, hit a ship and blew it up, the explosion usually did little damage, if any, to nearby ships. These explosions would typically send both large and small chunks of metal hurling through the air, but nearby ships would once again be protected by their Newtonian shields. The combined forces of air resistance and gravity forming the shields would cause the chucks to harmlessly fall into the ocean long before they could strike a nearby vessel.

Without air resistance, even the act of shooting at attacking aircraft would have been highly dangerous to other ships in a convoy. The 20-millimeter, 40-millimeter, and 5-inch (126 mm) cannon rounds fired by the thousands at approaching enemy aircraft were easily capable of inflicting serious damage to a navy ship if they struck it at full velocity. However, air resistance usually did a good enough job of slowing them down to prevent serious damage.

Typically, larger cannon rounds containing high explosives were fused to explode long before falling back to Earth. The shell would burst into hundreds of smaller misshapen pieces. Although still moving forward, they would be less aerodynamic. With higher air resistance, the pieces would quickly slow down. The lower velocity in combination with a smaller size would greatly decrease the damage if they rained down on nearby ships.

WWII antiaircraft gun crews wore helmets to protect themselves from such falling debris. While a helmet offered no protection from a cannon round at full velocity, it would protect against small pieces of falling shrapnel.

WWII-STYLE SPACE BATTLES

WWII naval battles have been the model for numerous movie space battles, such as those in *Star Wars*. Unfortunately, a good resistance force is hard to find in outer space, so fighting a space battle WWII-style would

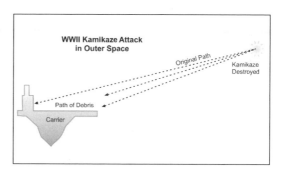

Figure 9: Kamikaze attack

be highly dangerous to all participants. Blowing up an enemy spacecraft, regardless of size, could damage other vessels, even at great distances.

If the fighter craft were small and attacking at high speed on a collision course with a much larger battle cruiser, there would be no safe way to "shoot down" the smaller craft. Such a craft could easily be closing in at several thousand miles per hour—many times the velocity of a speeding bullet. If the fighter craft were blown up, its pieces would still continue forward with many, if not all, pieces smashing into the battle cruiser. These pieces would not be like shrapnel falling from the sky. There would be no air resistance to slow them or gravity to pull them "downward." They would impact at high velocity and include not just large pieces of shrapnel but also gasses or plasma liberated by the explosion. These fluids could easily strike with velocities several times higher than the atomic blast wave at ground zero in Hiroshima (estimated at a mere 980 mph[4]).

Blowing up such a craft would be worse than turning a rifle shot into a shotgun blast. Okay, armor plating would work better against numerous small high-velocity particles than against a single large particle, but many of the smaller particles would gain kinetic energy from the explosion when the ship blew up. Even small high-energy particles would be capable of knocking out sensors and weakening armor plating, not to mention sending shock waves into a battle cruiser's interior.

If two large battle cruisers were flying side by side and a small spacecraft flew between them, the larger ships would face an added dilemma. If they shot at the small enemy ship and missed, they'd hit each other. Any projectiles they fired would go in a straight line. There would be no gravity to pull the projectiles downward into the water between them. (There would be no water either.)

A gunner could fuse his projectiles so they blew up in the space between the cruisers, but then he'd shower the opposite cruiser with high-velocity shrapnel. The fact that the shrapnel was misshapen and had poor aerodynamics would be no help. There's no air to provide air resistance in outer space.

Lasers and high-energy particle beams would have similar problems. There would be no way to limit a laser's range, and if one cruiser fired a beam and it missed the intended target, the beam would damage any other cruiser in the beam's path. But then such a beam should never miss. Even so, the problems of nearby exploding craft would persist.

Compared to space gunners, WWII antiaircraft gunners had far less danger of injury from blowing up nearby enemy craft, but they also had far more difficulty doing so. WWII projectiles had a downward curving trajectory and would have been moving slowly enough that a fast-moving aircraft could get out of their way. The WWII gunner couldn't aim directly at the aircraft. He would have had to estimate where the aircraft would be when the bullet arrived and how much the bullet would drop during its flight to the target—and then aim accordingly. Generally, thousands of rounds of ammunition were fired for every WWII aircraft shot down.

On the other hand, laser or high-energy particle beams travel in straight lines so fast that they arrive almost instantaneously. Even ordinary projectiles travel in straight lines and reach much higher average speeds in outer space, due to zero air resistance. It would be much easier for a space gunner to hit a target. The ratio of hits to beams fired would be vastly better than for WWII antiaircraft guns.

The difficulty of hitting aircraft with antiaircraft gun fire made bombers and torpedo planes formidable and generally reusable antiship weapons in WWII. If enough aircraft attacked from different directions, they could get through antiaircraft fire and sink a ship with bombs or torpedoes. If the pilots were willing to commit suicide, the chances were even better. In outer space, with the higher accuracy of laser or ultra-high-velocity particle beams, attacks of a smaller spacecraft against a larger one would invariably be suicide missions.

If the smaller craft survived the antispacecraft fire and used a torpedo, missile, or some other weapon to blow up a larger battle cruiser, the exploding cruiser would probably take its attacker with it. In fact, it would pepper all nearby spacecraft with high-velocity chunks of shrapnel. Blowing up a battle cruiser as depicted in movies would be a very dangerous thing to do.

While the WWII naval battle model used for Hollywood space battles is exciting, it's not realistic. So, how can Hollywood possibly justify its use? Filmmakers invented a convenient device called shields, which are supposedly force fields that surround space craft and protect them from harm. As long as the shields hold, one spacecraft can blow up another nearby and not have to worry about high-velocity shrapnel or blast waves.

It's arguable whether this is or isn't a legitimate use of artistic license. On the one hand, shields are an important plot device and are usually not overexplained with scientific mumbo-jumbo. In fact, they're rarely explained at all. On the other hand, there's no known mechanism to show how they could work, not even a reasonable theory.

The exception is a magnetic field used to shield against high-velocity charged-particle beams. A magnetic field will cause a force on a moving charged particle. Since the force will be perpendicular to the direction of motion, the particle will be deflected. This is how Earth's magnetic field helps protect us from the charged particles emitted by the Sun.

REALISTIC SPACE BATTLES?

In a "real" space battle—possibly an oxymoron—the craft would fight at great distance. If the ships were inhabited by biological beings similar to humans, these beings would have to breathe. At first glance, it seems that the object of battle would be to punching large holes in the hull to depressurize the opposing ship.

Aside from killing the inhabitants, the depressurization could disrupt weapon systems. The Joule-Thompson effect predicts that dropping the air pressure inside the ship would also lower its temperature. Certainly, if the temperature dropped to cryogenic levels (cold enough to liquefy or freeze most gases), the equipment would most likely cease to function. However, Joule-Thompson cooling alone would not be sufficient to do this.

Although we're used to thinking of space as a very cold place, there is no super-cold air in outer space to rush in when the ship depressurizes. It could take some time for the interior of the ship to reach cryogenic temperatures. After losing its atmosphere, further cooling would depend on radiant heat transfer. Radiant heat transfer tends to be slow unless an object is extremely hot, such as the surface of a star.

Since it would take a while to freeze the ship, a ship with robust automatic systems could continue fighting even after it was depressurized and all its biological inhabitants were dead. An enemy ship would have to be severely damaged or blown up to keep it from continuing the fight. That could be a challenge.

There are several ways an enemy ship could protect itself from attack. Beams of charged particles could be deflected with magnetic fields. Reflective coatings on the ship's hull or a cloud of reflective dust-sized particles surrounding the ship could help protect against lasers. Ordinary aluminized party balloons could be used to confuse incoming guided missiles. Since there's no air resistance, the balloons could be launched at high velocity. Put small thrust nozzles on the balloons along with some appropriate microcircuits and the gas pressure inside them could provide the thrust required for steering the balloons. The balloons could be programmed to behave like a flock of birds that would look like a spacecraft to distant attackers. Such a deception could draw fire away from the real ship.

Cloaking and stealth devices would be easy to implement in outer space. Paint the ship black, and at a distance it would be impossible to distinguish it from the blackness of outer space. Put a TV camera on one side and a TV screen the size of the ship on the other, and even the motion of the ship could be disguised. As the ship moved, the stars it blocked would be displayed on the big screen. Radar stealth could be achieved with the same technology currently used on Earth.

An entire book could be written on possible space battle weapons, decoys, defenses, and tactics. So why has Hollywood continued to use the WWII model instead of something more creative? The answer lies in what could be considered the first law of Hollywood inertia: once a movie proves profitable, any scene in it shall remain the standard of profitability until a movie with an alternative scene becomes even more profitable.

INDOOR GUN BATTLES

Blowing up spaceships in outer space is safe compared to the way the humans defend themselves in *The Matrix Revolutions* [NR] (2003) loading dock battle. (See Chapter 2 for a description.) Here, octopus-like sentinels stream toward defenders who are strapped to the front of robotic devices called APUs. With fully automatic 30-millimeter cannons attached to each APU arm, they blast away continuously at the sentinels streaming into the large concrete dome.

Unlike outer space, the room in the movie has two very dependable forces that can alter the forward motion of projectiles. The first is provided by the concrete dome, which, in the best case, stops cannon projectiles. If a collision with the dome doesn't stop a cannon projectile, it makes it change direction or ricochet. The second force, provided by gravity, then sends spent cannon shells back, along with concrete chunks, raining down on APU operators' heads.

Since distances inside the room are relatively short, air resistance has little effect because it doesn't have enough time to significantly slow down the ricocheting projectiles. Ultimately, the particles have to be slowed to a stop either by striking something they penetrate or by losing some velocity on each ricochet.

Every time they exploded, cannon projectiles containing high explosives would send high-velocity shrapnel and possibly concrete chips flying and ricocheting all over the room, increasing the probability of hitting people. Those unprotected by armor or helmets would be injured or killed.

Ironically, APU stands for armored personnel unit. Yet the personnel strapped to their fronts have no protective armor. They do not even wear helmets. This is one of the silliest battle scenes ever created. In reality, the APU operators would have killed or disabled each other in minutes from the combination of ricochets and falling debris.

The sentinels are equally suicidal (assuming the term can apply to machines). They form into streams and attack the APUs head-on. The APU operators merely have to stand their ground, point their cannons in one direction, and keep firing. However, the depiction once again ignores Newton's first law. Even a cannon shell will exert only a momentary resistance force as it tears through a sentinel. Like a kamikaze plane, the sentinel will continue forward and likely plow into the unprotected APU operator. Blow up a sentinel, and the pieces will pepper the APU with shrapnel.

It's not just the physics that make the scene ridiculous; it's the tactics. Why form a stream with hundreds of sentinels and flow straight toward the blazing barrel of a fully automatic cannon? Yes, that would overwhelm the APU operator thanks to Newton's first law, but it would also turn hundreds of sentinels into junk. If the sentinels had scattered and approached simultaneously from multiple directions, an APU operator couldn't possibly have swung and aimed his cannons fast enough to shoot down more than a few.

A more thoughtful group of sentinels would have bored a hole in the ceiling and dropped piñatas through it. When the APUs opened up, as mentioned before, they would have killed or disabled each other with ricocheting cannon rounds and falling bits of concrete. Imagine the scene: as the room falls silent, a bleeding APU operator reaches over and grasps the head of a broken piñata. Taking his last breath, he grimaces in the horrifying knowledge that he dies defending humanity from papier-mâché animals filled with toys and goodies—how poignant.

BULLET PENETRATION

Sometimes movies pretend there is a stopping force for projectiles when there isn't. In *Young Guns II* [PGP] (1990), Billy the Kid (Emilio Estevez) comes to the governor's house to bargain for a pardon. Always the shrewd negotiator, young Billy entertains the governor and his cronies with a little gunplay. The Kid blows the tips off several candles with rapid fire from his six-shooter, much to the delight of the group. Of course, this would have also blown holes through the window behind the candelabra and possibly killed hapless souls and livestock on the other side. As Newton's first law tells us, a bullet needs a force to stop its motion, and neither candles nor windows are up to the task.

Interior walls aren't much better than candles and windows for stopping bullets. In fact, with the exception of log cabins, even the exterior walls of wooden houses are usually inadequate. The common movie practice of hiding behind wooden walls while shooting at bad guys through the window would likely prove fatal. The bad guys' bullets would typically go right through the wall.

The Road to Perdition [GP] (2002) gets bullet penetration right. In the movie, Tom Hanks plays a 1930s salesman. He's a regular guy except for the .45 automatic he carries and the Thompson sub-machine gun he keeps in the garage with an extra fifty-round magazine of "sales literature" for those special occasions demanding rapid-fire "closings." Unfortunately, his son accidentally witnesses just such a closing, and Hanks is compelled to flee with his son in order to keep him alive. It seems that a sinister business associate doesn't trust the son to keep quiet.

After the associate hires a hit-man to kill the fugitives, Hanks fights back by stealing the organization's books from its accountant. With a shotgun under his coat, the hit-man sneaks up during the theft, and a gunfight ensues. The hit-man opens fire and blasts the wall behind Hanks, but Hanks prevails. He manages to wound the hit-man and escape. The buckshot goes through the wall and deposits itself in the accountant on the wall's other side. It seems like a trip to the hospital for a buckshot withdrawal would be in order, but alas, the hapless accountant has expired.

As simple as it is, one would think that Hollywood could get Newton's first law right; but they don't. In space battles, moviemakers neglect the dangers of having no resistance forces to slow down high-velocity projectiles. In *The Matrix Revolutions* loading dock battle, they ignore the peril of having forces that change the direction of high-velocity projectiles. In shootouts, they pretend that high-velocity projectiles can be stopped with inadequate forces. When they do get the forces right, it's worth noting. *The Road to Perdition* was not just paved with good intentions but with at least one good movie physics portrayal. By contrast, it seems that Hollywood's road to good physics isn't paved at all.

Summary of Movie Physics Rating Rubrics

The following is a summary of the key points discussed in this chapter that affect a movie's physics quality rating. These are ranked according to the seriousness of the problem. Minuses [–] rank from 1 to 3, 3 being the worst. However, when a movie gets something right that sets it apart, it gets the equivalent of a get-out-of-jail-free card. These are ranked with pluses [+] from 1 to 3, 3 being the best.

[–] [–] [–] Using the WWII naval battle model in space battles without thought about armor plating or shielding to protect spacecraft from other exploding vessels.

[–] [–] Using the WWII naval battle model in space battles and providing armor plating or shielding to protect spacecraft from other exploding vessels.

[–] Failing to account for the dangers of ricocheting bullets or bullet penetration.

[+] Depicting realistic bullet penetration.

[+] [+] Actually treating space battles in creative ways as though they are not simply extensions of WWII.

CHAPTER 6

NEWTON'S THIRD LAW:
That Special Hollywood Touch

NEWTON'S THIRD LAW—A SYNOPSIS

It's the yin and yang of physics, or as Paul Hewitt puts it, Newton's third law says that a person can't touch without being touched. Touch something forcefully, and it touches back with equal force.

Imagine a courtroom drama: the defendant stands accused of punching his victim in the nose. When asked for his plea, the defendant confidently replies, "Innocent." Why? When he punched the victim, the victim's nose punched back simultaneously with an equal but opposite force—not much of a defense, but true according to Newton's third law.

Boxers wear gloves as much to protect their hands from being broken as to protect the faces of their opponents from being cut. It's easy to break a knuckle when punching another person's head. The force on the knuckle is equal to the force on the head, and heads are often stronger. Movies have actually begun to recognize this fact. These days when a movie character throws a knockout punch, it's as likely as not that he'll grasp or shake his hand in an expression of pain.

Newton's third law tells us that forces always occur in pairs. Each force is exactly the same size or magnitude and occurs simultaneously but acts in the opposite direction of its twin. Force pairs always involve a pair of objects. One object creates a force on the other, and vice versa. If a moth smacks into the windshield of a bus, the force exerted on the moth by the bus's windshield is exactly the same magnitude as the force the bus exerts on the moth. Of course, the moth gets squished while the bus doesn't, but that's because it takes a lot less force to squish a moth than a bus.

HOW TO PREDICT MOTION—THE FREE-BODY DIAGRAM

If forces always come in pairs and are equal and opposite in direction, then how can physics predict motion?

The easy answer: the two forces act on two different bodies. Action-reaction force pairs can never cancel each other because they never act on the same object.
The fun answer: free-body diagrams (FBD). These show all the forces acting on an object from the outside world. By looking at a free-body diagram, it's clear whether the forces on an object cancel each other or not. If they don't, the object's motion is going to change. If the object is stationary, it's going to move.

The rules for FBDs:
1. Forces the outside world creates on the object are always shown.
2. Forces the object creates on the outside world are never shown.

3. Internal forces within the object are never shown.
4. Forces are always drawn touching the object.
5. Forces are always represented as arrows pointing in the correct orientation.
6. Other quantities such as velocity and acceleration are also represented as arrows and are sometimes drawn near FBDs but never drawn touching the object.

A picture of the object is usually simplified to a box or even at times a single point representing the object's center of mass. The example below represents a tennis ball struck by a tennis racket.

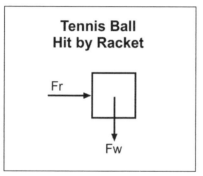

Where:
Fr = the force coming from the racket
Fw = the weight or gravity force acting on the racket.

Figure 11: tennis ball hit by racket

Note that even a round tennis ball can legitimately be represented as a square. The forces clearly show that the ball will be accelerated down and to the right.

The example below represents a stationary tennis ball sitting on the ground.

Where:
Fn = the normal force of the ground acting upward on the ball.

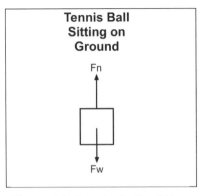

Figure 11: tennis ball sitting on ground

Note that all the forces cancel or counteract each other, indicating that the ball will remain stationary. Note also that, for convenience, the normal force (Fn) has been moved from the bottom to the top of the box. The normal force is the force the floor exerts on the ball, and although it acts on the lower surface, it's okay if the force is draw on the upper surface as long as the force's orientation with the vertical and horizontal dimensions is maintained.

SURVIVING WINDOWS

Billy Bob finds himself in possession of an old plate glass storm door and decides to demonstrate his manliness (based on a true story, well sort of). He sets it up in his back yard and, wearing the latest fashion in sleeveless t-shirts, rushes the window like it was the opposing team's quarterback, oblivious to the fact that plate glass doesn't like being "touched," let alone rushed. True to Newton's third law, the glass expresses its disdain by producing an equal and opposite "touching force" on Billy Bob, as it turns into hundreds of pointed, extremely sharp shards of glass. As Billy charges forward he pushes the shards horizontally out of the way—that is, except for the pointed ones that get stuck in his bare shoulder. The other shards slice as they are pushed aside. Many of them, pulled vertically downward by the force of gravity, slice and dice as they rain on Billy Bob, who is now a redneck with a blood-red neck.

Fortunately for Billy Bob, all the shards miss major arteries. While washing off the blood with his "hose pipe," he flashes a big bubba grin for his buddy's video camera. Yup, it does take the rest of the evening to pick out the glass imbedded in him, but hey, Billy Bob wasn't feeling any pain even before he charged the glass.

Very little force is needed for razor sharp glass to seriously cut a human. Jumping or walking through a plate glass window usually results in an injury—often a serious one. That's why safety glass was created. It's designed to shatter into small pieces with very few sharp edges. The smaller pieces reduce the amount of force an individual piece of glass can exert on the human it falls on and, in combination with the reduced sharpness, decreases the tendency for cutting.

Laminated safety glass adds a thin layer of plastic sandwiched between layers of safety glass for reinforcement. When the window is broken, the pieces tend to stick to the laminate rather than becoming deadly projectiles. Car windshields are generally made of laminated safety glass. Nevertheless, when craniums impact car windows at high speed, the result is often a head injury, including lacerations and broken bones.

Safety glass in general is four to ten times stronger (depending on whether or not it's laminated) than an equal thickness of ordinary glass. A large piece of it in a store window is a very hard surface and takes a fair amount of force to break. It also has a lot of inertia, so even when it does break, a lot of force is needed to move it out of the way. Naturally, the sheet of glass will create an equal and opposite force on the person who does the breaking and moving. While slamming into a large sheet of safety glass is far less likely to cause serious injuries than slamming into plate glass, it can still cause cuts, bruises, and broken bones, especially if done at high speed.

So, how does Hollywood send humans crashing through windows without so much as a scratch? The simple answer: they cheat.

At times, moviemakers have used panes made of sugar in glass-breaking scenes. That's right, candy windows! They look like glass and break like glass but have no sharp edges. More recently the candy has been replaced by a commercially available product called SMASH! plastic, which simulates glass without the safety problems. The product's manufacturer recommends that panes of the material be no more than 1/8 inch (3.2 mm) thick to avoid impact injury. The thin sheet reduces both the force required for breaking and the inertia of the fake glass.

Bottles used as clubs in fight scenes are generally made of fake glass, yet when a stunt man jumps through a window, it's often made of real glass. But fear not, there are still tricks involved. First, the glass will be safety glass. Second, when the stunt man runs toward the window, a helper blows the window out with small explosive charges an instant before the stunt man hits the window.

If the helper screws up and doesn't blow the window, the stunt man gets the vinegar knocked out of him when he hits the glass—not a good time to be the helper. Even when all goes well, stunt men sometimes still get cut. However, with safety glass and a correctly timed explosion, the cuts are usually minor.

In *True Lies* [PGP-13] (1994) Arnold Schwarzenegger is a secret agent who neglects to mention this fact to his wife. In one scene, Arnie has a fight and shootout with terrorists who attack him in the men's room. When one of the terrorists escapes, Arnie chases him through a shopping mall. The terrorist runs into a store and jumps through the display window onto the sidewalk beyond. When this scene is run in slow motion, it's easy to see the small explosive charges go off just before the stunt man hits the window. (Score one for the helper.)

NEWTON'S SCHOOL OF ACTING

Discharge a firearm, and a bullet is pushed forward. The gun recoils and is pushed backward exactly as predicted by Newton's third law (see Chapter 12 for calculations). However, fire a blank and very little force is needed to push the burning gunpowder out the end of the barrel because the powder has far less mass than a bullet, not to mention far less friction. Hence, blanks cause very little recoil. At first glance, it looks like the way to make realistic movies would be to use real ammunition. However, as mentioned in the last chapter, bullets have a nasty habit of going great distances, penetrating walls, and causing embarrassments like killing innocent bystanders. About the only solution to this is something called acting.

Actors need to understand some physics, especially if they play the roles of characters shooting firearms. Even though firing a blank produces little recoil, actors can make it look realistic. They need to spend a day being videotaped at a shooting range alternately firing real ammunition and blanks, and then adjust their fake recoil to match the real thing. They also need to try shooting from the hip in order to understand how ridiculous it is to do so (see Chapter 2).

The movie *Pearl Harbor* [PGP-13] (2001) (see Chapter 1) has one of the more egregious examples of absent recoil. During the Japanese attack, the dashing young fighter pilots Rafe (Ben Affleck) and Danny (Josh Hartnett) take to the air to fight back. With Zeros (Japanese warplanes) bearing down on their tails, they radio their ground crews and instruct them to "get some guns in the control tower." The ever-obedient ground crew all but fly up the stairs to the tower's top while carrying an assortment of weapons, including a .50 caliber machine gun weighing 84 pounds (38 kg).

They set the machine gun on a window sill and fire it, along with the other weapons, at a passing Zero. Of course they save the day and down the Zero with their very Hollywood-style mix of last-minute cleverness, enthusiasm, and teamwork.

Keep in mind that a single .50-caliber machine gun bullet is so powerful it can cut a moderate-sized tree in half with a single bullet. The recoil from a single shot would be nearly intolerable to a person firing it. Yet it spits out 550 rounds per minute. The weapon simply cannot be fired unless it's mounted on a specially designed tripod or a vehicle.

When mounted in a vehicle like a fighter aircraft, heavy machine guns will act like thrusters when they're fired. Although the recoil thrust is not enough to stop the aircraft, it will certainly jostle it about. The 20-millimeter cannons mounted in the wings of the Japanese Zeros would have produced an even greater recoil force per round fired than a .50-caliber machine gun. When firing these cannons, the aircraft would have vibrated significantly, causing small changes in the aim point. Add to this the normal bumpiness of flight, and the cannon shells would have scattered in a random manner around the point of aim. So how do the moviemakers in *Pearl Harbor* portray strafing runs by cannon-firing Zeros? They have the cannon shells strike the water in evenly spaced rows lining up perfectly with the cannon barrels in the wings.

NEWTON'S THIRD, AT A DISTANCE

Newton's third law works at a distance as well as up close. When someone jumps out of an airplane (hopefully with a parachute) Earth exerts a downward gravitational force on the person, and the person exerts an upward gravitational force on Earth. The two forces are equal in magnitude but opposite in direction. Of course, the skydiver does almost all of the moving. She has a much lower mass and hence a much lower inertia than Earth. Inertia, as mentioned earlier, is resistance to motion.

In the famous fight between Yoda and Count Dooku (Christopher Lee) in *Star Wars Episode II* [NR] (2002), Dooku uses the Force and causes large chunks of the ceiling to fall. About to be crushed, Yoda raises his palm and projects a force that stops the heavy pieces in mid air. To do so, Yoda must maintain an upward force on the rocks equal to their weight. Newton's third law says that the rock will simultaneously exert an equal but opposite force on Yoda, not a healthy situation for the little guy. Yet, Yoda is uninjured.

Yoda had intervened to prevent Dooku from killing both Anakin Skywalker (Hayden Christensen) and Obi-Wan Kenobi (Ewan McGregor). Dooku had first disabled Obi-Wan, who was left lying helpless on the floor. When Anakin came to the rescue, Dooku lopped off his arm. Raising his palm, Dooku projected a force and sent Anakin flying. Of course, Newton's third law says that Anakin would have created an equal but opposite force on

Dooku. Yet Dooku shows no sign of being pushed backwards. Even his raised palm and arm show no signs of recoil. Evidently, the force he projected was not just from a galaxy far, far away but from a different universe—one that doesn't follow Newton's third law.

NEWTONIAN MARTIAL ARTS

By using the right technique, it is possible to remain stationary while pushing a person backwards. If the push is applied with a slightly upward motion near the center of the pushed person's mass, it tends to lift him slightly off his feet and move him backward. A forceful push in this manner can so seriously disrupt a person's balance that he is sent running backward across an entire room. It looks incredibly fake, but it's not! The initial push sets the person in motion, causing him to take a step backward to keep from falling. Unfortunately, the step reinforces the backward movement, which requires yet another step and another and another in rapid succession to keep from falling down. Often the person has to run into something, such as a wall, in order to stop.

An equal but opposite force acts on the person doing the pushing. However, this force acts slightly downward as well as backward. If the person doing the pushing is relaxed and standing in a stable martial arts stance, she will not lose her balance and be thrown backward. The slightly downward direction of the reaction force acting on her tends to push her feet more firmly against the ground, which helps hold her in place.

The pushing action described above is commonly done during pushing hands practice in the martial art of Taijiquan. Pushing hands is a type of sparring in which the participants stand in fixed positions and try to unbalance each other.

Participants are expected to remain relaxed and generally use very little physical force. It's definitely not a Western-style wrestling match.

Taijiquan is often practiced for its stress-relieving effect and is notable for its slowly flowing individual forms. It's also an effective martial art. In some ways it's similar to Yoda's use of "the Force." Taijiquan practitioners visualize chi or life force circulating through their bodies that can then be used for performing what seem like superhuman martial arts feats. In reality, they are merely remarkable applications of simple physics.

Taiwanese moviemaker Ang Lee got the physics right in his gemstone, *Pushing Hands* [GP] (1992). In the movie Mr. Chu (Sihung Lung) is a Taijiquan master who has not only lost his wife but been forced to move to America and live with his son. His daughter-in-law has typical yuppie values that clash with the traditional ways of Mr. Chu. Admirably, Mr. Chu spends a significant part of his day watching and criticizing the ridiculous martial arts depictions in Hong Kong kung fu movies. Thanks to his martial arts expertise, he is unable to suspend his disbelief while watching such nonsense.

In one scene Mr. Chu is teaching pushing hands to a group of students in a large room. At the other end of the room, the widow Mrs. Chen (Jean Kou Chang) is teaching a cooking class. Mr. Chu wants to get to know her but is not one for the usual shallow pickup lines. Instead, he begins practicing pushing hands with a rotund student and carefully lines him up with Mrs. Chen's table. At just the right moment, Chu pushes his hapless student and sends him running backwards across the entire room while trying to regain his balance. The student crashes into Mrs. Chen's table, bringing about the meeting.

In the *Star Wars* movie, Dooku's push is far more forceful than Mr. Chu's and looks like it is directed slightly downward rather than upward, judging by the position of his palm. Anakin flies completely off his feet, so his backward motion has nothing to do with losing his balance. Dooku should have recoiled backward and upward, but as mentioned before shows no sign of it.

If a Taijiquan master applies a larger force with a more upwardly directed angle than used in pushing hands, he can also send a person flying, but it's just a few feet off the ground and a few feet backward. It's nothing like the exaggerated backward flight of Dooku's victims. Pushing people off their feet is typically not done in pushing hands practice, for the obvious reason that it can cause injuries. A push with even greater force will act like a palm strike rather than a push. Here the result is more likely to be broken bones or internal injuries than a dramatic backward motion. The human body simply cannot hold up to the force required to send it flying across the room with a quick push.

So how do special effects experts send actors flying across the room in movies without injuring them? First, they fit the actor with a specially designed vest to distribute the force over as much area as possible. A rope or wire is attached to the vest on the person's back, and at the right moment the person is pulled slightly upward and backward, making him or her fly across the room. It takes a much lower force to do so since it is applied for a longer time.

NEWTON IN SPACE

The movie *2001: A Space Odyssey* [GP] (1968) is often cited as one of the best examples of movie physics, yet it too contains some questionable Newton's third law scenes.

In the movie an out-of-control computer named HAL controls every aspect of a spacecraft on a mission to Jupiter. When a pair of humans questions his decisions, HAL decides to deactivate them. He waits until one is attempting a repair during an extravehicular activity (EVA), then attacks using the mechanical claw on a nearby space pod, leaving the hapless human adrift in space, with a broken air hose, struggling for his last breath. The other human, Dave, immediately attempts a rescue using the remaining pod. This proves futile, and Dave ends up begging HAL to open the mother ship's airlock door so he can come back inside. Surprise, surprise—HAL refuses.

Naturally, there's an emergency airlock that can be opened from outside, but there's also a small problem: in his haste, Dave forgot his pressurized helmet, and he cannot dock the pod with the airlock door. To use the airlock he must first exit the pod into the vacuum of outer space. His exposed head will be subjected to the vacuum of outer space before he can enter, close the door, and pressurize the airlock.

He uses the pod's claw and opens the emergency airlock anyway. He then lines up the pod's door with the airlock, holds his breath, and blows the door's bolts. Conveniently, these have been designed with explosive devices built into them. We know this because it's written on the outside of the pod.

Escape hatches with explosive bolts were standard equipment on capsules used in early NASA manned space flights. The capsules returned to Earth by parachuting into the ocean. If one started sinking, it was handy to have a quick way out. Why a fast exit is needed in outer space is a mystery.

After blowing off the door, Dave shoots into the airlock, presumably propelled by air pressure escaping from the pod. He bounces around and almost immediately closes the airlock door. He survives, and HAL is now at his mercy.

When the door, Dave, and the air were expelled out the back of the pod, the pod should have gone flying the other direction into space but did not. Dave may have set the pod's thruster controls so that it would push the pod against the mother ship; however, it's unlikely that the thrusters would have been strong enough to completely counteract the reaction force created on the pod when the door blew. It's a flaw but is perhaps forgivable in light of the many things the movie *2001* did right.

Even a prestigious scientific organization like NASA has had problems with Newton's third law. Before attempting to send people to the Moon, NASA thought it wise to work out a few details such as space-walking or EVA[5]. Astronauts might need to go outside their spacecraft and make repairs on long journeys to the Moon, just as depicted in *2001*. Unfortunately, when they tried EVAs during the Gemini program, they received some nasty surprises due to Newton's third law.

The first EVA went flawlessly, but the astronaut involved, Ed White, did not attempt to do any work. The second EVA was a near disaster: Gene Cernan was supposed to venture out of the spacecraft and put on a flying backpack while on the dark side of Earth, where he had very little light. Every time he attempted to turn a valve, his entire body turned. Anything he touched touched back and repelled him. He had neither gravity nor friction to hold him in place.

After just a few minutes, Cernan began to overheat and sweat from exertion. His space suit became like a steam bath. His heart raced at 170 beats per minute. Inside the Gemini capsule, fellow astronaut Tom Stafford became increasingly concerned. He knew that if Cernan lost consciousness, there would be no way to get him back in the capsule for reentry. Stafford would have no choice except to cut Cernan free and leave him floating in orbit. In consideration of the desperate situation, the EVA was aborted. Both men returned safely to Earth, but when NASA workers examined Cernan's space suit, they poured over a pound and a half of sweat out of each boot.

The next two EVAs went almost as badly as Cernan's. Finally, after three bungled tries, NASA started thinking about how Newton's third law worked. Subsequently, they completely reworked their EVA training, procedures, and equipment. After these changes, EVAs became routine. When NASA has a flaw in its understanding of a situation's physics, it gets fixed fast. By contrast, Hollywood serves up the same physics mistakes over and over again. It could do better if revenues were any indication of its ability to do so. Hollywood gleans about nine billion dollars a year from newly released films alone[6], not including rentals, DVDs, videos, and spin-off product royalties. By comparison, NASA's Gemini program cost only 1.28 billion dollars over three years, with only a small fraction used for fixing the EVA flaws related to Newton's third law. However, a flaw has to be defined as a problem before it can be fixed, and here the difference between NASA and Hollywood is striking. For NASA problems are defined in life-and-death terms. Since even minor flaws in understanding a situation's physics can be deadly, they are, by definition, problems. For Hollywood problems are defined in terms of profit. When a movie with egregious physics flaws turns a profit, by definition it has no problems.

Summary of Movie Physics Rating Rubrics

The following is a summary of the key points discussed in this chapter that affect a movie's physics quality rating. These are ranked according to the seriousness of the problem. Minuses [–] rank from 1 to 3, 3 being the worst. However, when a movie gets something right that sets it apart, it gets the equivalent of a get-out-of-jail-free card. These are ranked with pluses [+] from 1 to 3, 3 being the best

[–] [–] Actors slamming their fists through car windows with no discernible injury.

[–] [–] Actors jumping through plate glass windows with no discernible injuries. (Note: plate glass would be found in older buildings or in ordinary residential windows. Sliding glass doors are an exception. They usually are made of safety glass.)

[–] Actors effortlessly jumping through large safety-glass display windows.

[–] Actors shooting high-powered firearms without any discernible recoil.

[–] The bullets from a strafing aircraft striking in nice neat evenly spaced rows.

CHAPTER 7

CREATIVE KINEMATICS:
Explosive Entertainment

WORLDWIDE EXPLOSIONS

An ancient asteroid impact triggers a sea of flame spreading about 6,000 miles (9,700 km) around the globe in seconds. It brings fiery death to everything in its path: trees, ferns, insects, and dinosaurs (opening scene *Armageddon* [RP]). Modern evidence indicates an ancient impact did indeed touch off a major firestorm and doomed the dinosaurs, but probably didn't look like the movie version. First, there's the flame-front speed depicted in the movie, calculated as follows:

SPEED = DISTANCE/TIME (EQUATION 7.1)

= 6,000 mi / (30 sec 1/3600 hr/sec)
= 720,000 mph or 1,160,000 kph

This speed is about 29 times faster than the minimum speed needed for an object to escape Earth's gravity (escape velocity of Earth) and 950 times faster than the speed of sound: fast enough to blow away Earth's atmosphere—impossibly fast.

A global-sized flame front spreading at a speed greater than the speed of sound is by definition a detonation. To maintain such velocities, the Earth would just about have to be covered in a thin sheet of explosive material like TNT. A rapidly expanding fire ball is like a jet airplane: even when it runs out of fuel, it will move forward, but not for hundreds and hundreds of miles. To keep moving it must have a source of fuel and the ability to consume it fast enough to maintain its speed, hence the need for the TNT. Even a flame front traveling at the edge of subsonic speed (just below 760 mph or 1200 km/hr) would have to consume combustibles in its path at an explosive rate to fuel its high velocity. Such high velocities could not be maintained without continual energy input.

This near-sonic-speed wall of fire would take about eight hours to spread 6,000 miles around the world. While such a front might be hundreds of miles wide, it would burn so fast that it would leave a large burned-out area behind it as it spread. From space, the flames would look more like a wide slow-moving ribbon than an all encompassing sea of fire—that is, if the flames were not completely obscured by smoke.

At the other extreme, forest fires release the energy stored in brush and trees much more slowly and generally travel at less than 4 mph (6.4 km/hr)—a speed that would take more than two months to spread 6,000 miles around the world. By either estimate—high or low—an entire hemisphere is unlikely to be engulfed in flame all at the same time.

Still *Armageddon's* opening depiction of an asteroid strike earns reasonably good marks. Its extraordinarily high flame velocity can be forgiven as a time-lapse effect, its sea of flame as over exuberance. The depiction could have benefited from scientific studies done with computer simulations but the overall visual effect was scary enough to make even a politician think about preventative action—a much needed activity. Unfortunately, the rest of the movie had a solution about as reliable as a campaign promise.

A realistic defense for preventing an asteroid strike could take decades and billions of dollars to develop. Sadly, it will probably also take a disaster—hopefully one in a lightly populated area—before humanity is willing to spend the money. Splitting a major sized asteroid in half with a nuclear bomb from our cold war arsenal will not be part of the plan. Such bombs simply do not have the required energy output (see Chapter 11).

How Explosions Propagate

An explosion's effects are related to the ways its energy propagates, or spreads. As a rule of thumb these are:

1. Blast front: a wind-like movement of expelled materials including gasses, plasma, or debris such as shrapnel. Expelled materials can travel at very high, even supersonic, initial velocities. This material usually includes any fireball from fuel or explosive not consumed in the initial blast. Expelled material is often superheated and can cause secondary fires and burns.

    On Earth, if expelled gasses travel at supersonic speeds they can compress air ahead of them enough to cause superheating and result in secondary fires. Expelled materials lose kinetic energy rapidly due to air resistance or, in the case of solid debris such as shrapnel, due to being pulled to the ground by gravity. In outer space expelled materials generally lose kinetic energy only when they impact another object.

2. Shock wave: a high pressure pulse traveling as a wave through air and caused by the blast front moving at or above the speed of sound. The shock wave can continue traveling considerable distances at the speed of sound long after the blast front has slowed down.

   On Earth shock waves can do considerable damage. In outer space there are no shock waves because there is no matter to act as a medium for propagating the wave.

3. Electromagnetic (EM) pulse: a broad spectrum pulse of electromagnetic energy that can include everything from radio waves to gamma rays (for nuclear blasts). The pulse travels at the speed of light and can interfere with electronic equipment. It often includes a large amount of infrared radiation, enhanced by fireballs, burning objects, or superheated materials emitting infrared or thermal radiation for as long as they remain at elevated temperatures. This radiation can set secondary fires and cause burns.

In outer space, the blast front and EM pulse expand like two giant bubbles—surface area increasing with the square of distance from the blast—albeit the EM "bubble" expands much faster than the blast front. Since both kinetic and EM energies remain constant the intensity of the explosion (the total energy absorbed per unit of area for structures in the path of the blast) will decrease with the square of the distance from the blast. Doubling distance reduces intensity to one-fourth its original strength. The exceptions are chunks of solid debris or shrapnel which can be just as damaging at a distance as up close. On Earth, an explosion's intensity will decline even faster due to air resistance. In general, an explosion in space can damage at much greater distances than its equivalent on Earth.

Armageddon concluded when the Texas-sized asteroid, on a collision course with Earth, was split in half by—you guessed it—a nuclear bomb, just in the nick of time to save humanity. The plume from the blast radiated outward in the shape of a disk about three times the diameter of the asteroid (a total comparable to the distance across the United States) in about two seconds (a speed of about 4,000,000 mph or 6,400,000 kph)—quite a blast for a device that normally produces a fireball a few miles in diameter and a shock wave traveling no more than a few times the speed of sound (760 mph or 1200 kph).

EXPLODING WORLDS

Hollywood recipes for planetary disasters are not just served with baloney, they're made of it. When the Empire's Death Star blows up Alderon (an Earth-sized planet) the pieces fly outward, amid a swirling orange-white fireball, a distance of about twice the diameter of the planet in about two seconds—a speed of 28,800,000 mph (46,400,000 km/hr), all the more remarkable because the exploding particles have to overcome the gravitational forces pulling the parts back into the form of a planet. Keep in mind that a typical detonation travels no more than a few times faster than the speed of sound (760 mph or 1200 km/hr). While there is no law of physics that says a planet can't be blown apart at such high velocities, certainly the numbers cast doubt on the notion that a death star could do it.

Okay, blowing up an entire planet is unlikely but if it did happen, would it look like the *Star Wars* depiction? Certainly the explosion would be like a gigantic nuclear blast and as a rule of thumb, about half of the energy in such a blast goes into heat and the rest into kinetic energy. The kinetic energy of a single 1 kg blob moving at the speed depicted in the movie would be about the equivalent of 20 kilotons of TNT, not quite twice the energy released by the 12.5 kiloton bomb dropped on Hiroshima. An equivalent amount of heat would be enough to vaporize a city, let alone a 1 kg blob—duh. Multiply the amount of energy for a single blob by the 6×10^{24} (more than a trillion times a trillion) similar blobs contained in an Earth-sized planet and the amount of energy released in the blast would be the equivalent of the Sun's total energy output for about 800 centuries—released in a couple of seconds. Would the blast look like the movie depiction? Not likely. The planet would vaporize in a huge flash of blinding light. When the flash began fading, the vapor would start condensing into a gigantic slowly expanding dust cloud.

A less extreme exploding Earth-like planet would probably look like a balloon filling with liquid to its limits then popping, all in slow motion. The liquid in this case would be the glowing molten material of the inner planet. To fly apart the pieces would have to travel at least at terminal velocity (25,000 mph or 40,300 kph for Earth). Assuming a speed twice as high as terminal velocity, the debris from the planet's surface would expand outward a distance of twice the planet's diameter in about 19 minutes. On the other hand, if the pieces didn't reach terminal velocity, they would fly outward and then collapse backwards into a turbulently swirling molten planetary blob. Since the mass contained in the planet's solid crust would be small compared to the planet's molten interior (assuming it's Earth-like), the crust would essentially be swallowed up in a sea of brightly glowing lava. While not as dramatic as a sudden explosion, a more realistic explosion would have its own type of horror: the type that comes from the slow realization that a catastrophe is occurring and there's nothing that can be done to stop it.

Filmmakers are well aware that blowing up a small scale model does not look the same as blowing up the real thing. Small explosions have similar velocities to large ones, but the flying pieces travel much shorter distances making the small explosions appear to happen much faster than the large ones being simulated. Cameramen compensate with slow motion photography. They film small explosions at higher than normal frame rates so that when the film is projected at its normal rate, the explosions are slowed down and look like they're full sized.

Filmmakers should use the same principles in reverse to deduce that a planetary explosion—occurring on a gigantic scale—would look like it was happening in slow motion. Movies with exploding planets are rare, but it's safe to say that the next one will probably look like the last one. It's the law of Hollywood inertia: never alter the formula used in a successful movie.

THE SUPER SPEED OF SPACE TRAVEL

Space travel poses a different speed problem: it takes an incredibly high speed to get anywhere. The Apollo 10 mission holds the speed record for manned spacecraft at roughly 25,000 mph (40,300 km/hr). Double it and it would still take about 13 years to travel across the solar system (assuming that it is roughly circular with a diameter equal to the average distance from the Sun to Pluto of 2.8 billion miles). Travel to the nearest star outside our solar system, Proxima Centauri at a distance of 4.3 light-years or 2.5×10^{13} miles (4.1×10^{13} km) would take about 58,000 years—a little long to keep the kids in the back seat alive let alone entertained.

Decide to travel around our galaxy (the Milky Way) and the need for speed gets even more extreme. Traveling at the speed of light—about 1,300 times faster than the 50,000 mph used in the previous examples—it takes about 100,000 years to travel across the galaxy. To go much of anywhere, a spacecraft would need to travel about 1,000 times the speed of light, assuming that human lifetimes can be expanded to at least 200 years and suspended animation technology is available to facilitate return trips. But above all else, society would need the willpower to devote the major resources required for such journeys. Currently, we can't even cough up the funds for a mere moon base. Of course, inter-galactic travel could be done more cheaply with machines than humans, but there's no movie in that.

Unfortunately, the speed limit for spacecraft is set below the speed of light, at least according to the famous galactic traffic cop, Albert Einstein. On a practical basis, it's set far lower.

Einstein taught us that the mass of an object approaches infinity as the object approaches the speed of light—a puzzling statement that makes a lot more sense if the word inertia is substituted for mass. Recall: inertia is resistance to change in motion. Einstein is saying that it gets more and more difficult to change an object's motion once it reaches speeds near the speed of light—making it infinitely difficult to actually reach such a speed. And infinitely difficult problems are rather hard to solve. To put the problem in perspective: at 25 percent of the speed of light (c) the inertia is 3 percent higher; at 50 percent c, the inertia is 15 percent higher; at 99 percent c, the inertia is 709 percent higher than at rest. Obviously, there's no way to carry enough fuel to reach the speed of light.

The problem is further complicated by something called time dilation—exemplified by the twin paradox. Find a set of twenty-year-old twins, leave one on Earth, and send the other into space on a lengthy trip at a speed of 99 percent c. On returning, if she has aged by ten years, her twin sister—who stayed on Earth—will have aged by seventy one years. Start sending star ships on long missions at similar speeds all over the galaxy and no one will be able to keep their clocks synchronized. The result: a major breakdown in organized exploration.

Star Trek solved the problem for its spacecraft by warping space. Take a sheet of paper and ask a friendly ant to walk across it as you time the journey. It will take a while. Fold the paper so that it hangs in a loop with the ends touching each other at the top of the loop; the ant will need much less time to traverse from one end to the other. You have just warped space—as far as the ant is concerned. No one knows how to do it for a spacecraft but at least it's conceivable.

Star Trek's space ships can travel the galaxy at sublight speeds, keep their clocks synchronized, yet warp space for quick long distance journeys. According to Lawrence M. Krauss in *The Physics of Star Trek* there is even some theoretical support for it. Space could conceivably be warped by a super strong gravitational field. But don't hold your breath waiting for the day. There are monumental problems in the way. As for more conventional thruster type technology, even short hops around a solar system will continue to be expensive, lengthy, and difficult.

OUTRUNNING EXPLOSIONS

Not all speed problems are galactic: take the problem of outrunning fiery explosions—a useful skill on the human scale. Sometimes the explosion occurs in the open and the object is to run and jump into the nearest body of water before the fireball hits the would-be escapist. Once underwater, the camera generally shows the deadly fireball sweeping overhead. If the explosion were far enough away—say, 1000 meters away—such an escape might actually be possible.

THE PHYSICS OF OUTRUNNING WHATEVER

When someone tries to outrun something like a car or an explosion he or she has to have a head start or it's hopeless. To model the situation we'll assume that the subject or person running away, as well as the object pursuing, both move at constant velocity. We'll also assume that the object being outrun moves faster than the person. Otherwise, the subject is in no danger of being overtaken, and what drama is there in that? We'll represent the head start as a distance dh. Distances can be calculated from the following kinematic equation:

$d = v \bullet t$ (EQUATION 7.2)

Where:

d = distance (note this is actually displacement or distance in a straight line)

v = velocity

t = time

We'll use o and p subscripts to denote the object doing the chasing and the person chased respectively. The distance traveled by the object doing the chasing looks like this:

$d_o = d_p + d_h$ (EQUATION 7.3)

$$\text{but, } d_o = v_o \bullet t$$
$$\text{and, } d_p = v_p \bullet t$$

Substituting the two above equations into equation 7.3 yields:

$(v_o \bullet t) = (v_p \bullet t) + d_h$

$(v_o \bullet t) - (v_p \bullet t) = d_h$

$t (v_o - v_p) = d_h$

$t = d_h / (v_o - v_p)$ (EQUATION 7.4)

Equation 7.4, as mentioned above, will work for any type of chase situation including either an expanding fireball or car chasing a person attempting to run away. For example, assume an explosion is 50 meters away traveling at the speed of sound and the person running from it can run 100 meters in 10 seconds (9.1 m/s, fast by any standard). The time such a person has to escape before being engulfed by the fireball is calculated as follows:

$$t = 50 \text{ m} /(340 \text{ m/s} - 9.1 \text{ m/s})$$
$$t = 0.15 \text{ s}$$

The distance the person could run is calculated as follows:

$$d_p = 9.1 \text{ m/s} \ (0.15 \text{ s})$$
$$= 1.4 \text{ m or } 4.6 \text{ ft}$$

If the person were standing on the edge of a dock and dove off a dock exactly 0.15 seconds before the fireball arrived, the distance he or she would fall toward the water would be calculated as follows (assuming that the person could not push off the dock in a way which gave an initial velocity in the downward direction):

$$d_y = \tfrac{1}{2} \ gt^2$$

Where:
d_y = the vertical distance fallen
g = the acceleration due to gravity. (On Earth g = 9.8 m/s2)

$$d_y = \tfrac{1}{2} \ (9.8 \text{ m/s}^2) \ (0.15 \text{ s})^2$$
$$= 0.11 \text{ m or } 4.3 \text{ inches}$$

Obviously, the person is going to get flamed before he or she hits the water. But, not all fireballs travel at the same velocity. If the fireball traveled at a much more sedate speed of say half the speed of sound, the person attempting escape would have a whopping 0.3 seconds to escape and could fall 17.2 in (44 cm) toward the water—a short enough distance to still get torched.

Within 50 m of the explosion, victims are pretty much doomed to be engulfed by the fireball (assuming it's big enough) and slammed by the shock wave. Even in the best case scenario, with a "slow" moving fireball, victims will have at most a few tenths of a second to run and jump in the nearest body of water. If they are neither knocked unconscious nor killed outright by the shock wave, flying debris, or shrapnel; and they hit the water soon enough to avoid severe burns, they might even survive.

The probability of survival without horrible injuries increases rapidly with increasing distance from the blast. But the benefits of running and jumping in water remain limited. Fireballs with enough energy to travel great distances usually do so at high velocities. Such enormous high-temperature fireballs also emit large amounts of infrared radiation (IR) that can burn victims at a distance even when the fireball does not contact them. The IR radiation travels at the speed of light—a little hard to outrun. If a person is close enough to the blast to be in grave danger, chances are he or she will not have the time to run and jump in the water. On the other hand, if a person has the time to run and jump in the water, chances are he or she is too far away from the blast to be in any real danger.

So how do movies do it? The actor stands in front of the fuel dump, building, or whatever is to be blown up and is filmed using a telephoto lens. This makes the actor appear to be standing close to the object to be blown up when in reality he or she is at a safe distance. The camera is turned off, a stunt person is substituted for the actor, the camera is turned back on, and ka-boom. The explosion is started with a black powder blast demolishing a container of fuel, the fuel mixes with air, and a second black powder explosion is used to ignite the vapor into a fireball. The fireball's image fills the entire screen making it look enormous. Black powder in combination with a fuel is ideal because its explosion produces lots of highly visible smoke and flame with relatively little damaging blast power (compared to more powerful explosives like dynamite, TNT, or C-4, which give off very little smoke and flame with powerful blasts). Energy used to produce smoke and flame is essentially wasted because it's unavailable for producing the high-pressure blasts that pulverize materials. The more powerful explosives are less visible and less spectacular precisely because they are more efficient and effective for just about every purpose but movies.

Saving Private Ryan and the miniseries *Band of Brothers* did an excellent job of depicting explosions. Neither artillery shells nor hand grenades produce large fireballs in these movies. In real life, even grenades such as those containing white phosphorus or thermite designed for marking targets or starting fires do not produce the typical fireballs of burning gasoline seen in movies. An exploding white phosphorus grenade looks very similar to a white fireworks shell bursting on the ground and sending glowing streamers flying outward in every direction. Although white phosphorous burns at about 5000°F (2760°C), a white phosphorus grenade does not produce a large-sized yellow-orange fireball. Thermite grenades typically are not even designed to burst. They burn vigorously in a local area at temperatures of around 4000°F (2200°C) and produce a by-product of molten iron. These grenades work extremely well for destroying enemy equipment such as artillery pieces. By contrast, the commonly used general purpose fragmentation grenade produces even less fire and smoke. It's designed to convert the grenade's explosive energy into the kinetic energy of hundreds of pieces of high velocity shrapnel. For such a grenade, fire and smoke are a waste of energy.

EXPLOSIONS IN SHAFTS AND TUNNELS

There are many variations of the running from explosions theme, such as the fiery elevator shaft. The heroes are climbing out of the elevator shaft as a fireball races up from below. They pull themselves out just as the flames sweep past—not a problem, they're heroes. Long tubes such as elevator shafts confine fiery blasts and increase their pressure thereby increasing their velocity. Yet, heroes—in moments of great stress—have that unique ability to tap into their simian ancestry of millions of years past and conjure up long dormant genetic abilities to climb out of elevator shafts hundreds of times faster than competitive runners can run.

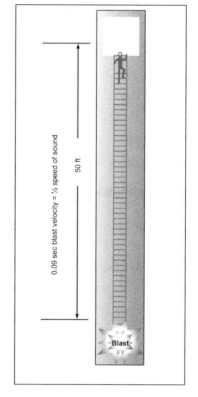

Figure 13: blast velocity

There are also the escapes in which a hero on a motorcycle skillfully outruns a gigantic blast-wave/fireball. If he's 100 m from the blast and going 100 mph (161 kph), he's got 0.33 seconds to escape. This is naturally enough time for him to glance backwards repeatedly at the impending doom and lay down the motor cycle just in time so he ends up in a convenient ditch as the blast passes overhead—all with no injuries.

Collateral Damage [PGP-13] combined the fiery explosion in a tunnel and the motorcycle escape into a single scene when a pair of villains on a motorcycle set off a natural gas explosion, racing down a tunnel while firing a handgun at the movie's hero, Arnold Schwarzenegger. Arnie, naturally, not only dodges the bullets but also outruns the blast and ducks behind a door just before the explosion hits. The less fortunate villains are knocked off their motorcycle by the blast (they are, after all, villains and aren't expected to outrun it). While bruised and battered, they, nevertheless, are able to engage in lengthy hand-to-hand combat with Arnie who eventually proves to be far more deadly than the explosion. This illustrates yet another law of Hollywood: really evil villains can never die from the first few fatal causes.

This chapter has been able to debunk numerous typical Hollywood movie scenes with a single basic kinematic equation: the equation for constant velocity, which is often taught in one class period at the beginning of high school physical science—not exactly rocket science. So why do moviemakers assume that their audiences won't notice? As described in Chapter 1, they're counting on the power of the movies: combine rousing music with dramatic images in adrenaline-pumping scenes, and even silliness can go straight into the brain unopposed by logic.

Summary of Movie Physics Rating Rubrics

The following is a summary of the key points discussed in this chapter that affect a movie's physics quality. These are ranked according to the seriousness of the problem. Minuses [-] rank from 1 to 3 with 3 the worst. However, when a movie gets something right that sets it apart, it gets the equivalent of a get-out-of-jail-free card. These are ranked with pluses [+] from 1 to 3 with 3 being the best.

[-] [-]   Planets that explode in a few seconds.

[-] [-]   Heroes outrunning or outclimbing nearby fiery explosions, especially those in elevator shafts or tunnels.

[-] [-]   Traveling the galaxy or even around a large-sized solar system with spacecraft powered by conventional thrusters.

[-] [-]   Fragmentation hand grenades or high explosives such as TNT, C4, or dynamite detonating with large fireballs.

[0]   Terrestrial fireballs traveling great distances at hypersonic speeds (incorrect but forgivable).

[+] [+]   Fragmentation hand grenades or high explosives such as TNT, C4, or dynamite detonating without large fireballs.

CHAPTER 8

HOLLYWOOD BOMBS:
How Filmmaker Physics Misses the Boat

THE PHYSICS OF BOMBING

Sailors sleep as the aircraft approaches, its bombardier squinting through his bombsight at the toy-like image 9,840 feet (3000 m) below. When the aircraft is directly over the target below, the bombardier releases his deadly payload. It falls straight down, penetrating deep into the USS Arizona and blasting it into history as the emblem for the United States' worst military defeat. No, it's not the day of infamy. It's the movie *Pearl Harbor* [PGP-13] (2001) perpetuating the infamous physics misconception that bombs dropped from moving aircraft fall straight down.

If all the Japanese bomber crews had shown such ignorance of physics, none of their bombs would have hit their targets and the USS Arizona might today be a floating museum rather than a sunken tomb. At best—or worst, depending on whose flag one saluted—only Japanese torpedo planes would have damaged ships during the first-wave attack.

Had the attack been planned by physics fools, even the torpedoes would have gone awry. Pearl Harbor was notoriously shallow, so torpedoes dropped from aircraft had to be modified with wooden fins and dropped from a carefully determined height to keep them from going too deep and slamming into the muddy bottom. Errors in understanding the physics would have rendered the torpedoes useless.

Dive bombers might have remained effective even in a time of physics foolishness, but then they didn't attack ships—at least, not in the first wave. Dive bombers used what could be called bombing physics for dummies or, more correctly, bombing physics for maniacs. Start from the same altitude as high-level bombers, dive straight at the target, release the bomb at maniacally close range, then pull up sharply to avoid becoming as one with the target—hopefully without blacking out from multiple gs. Unlike a bomb dropped from a high level that depended only on gravity for downward velocity, a dive bomber's load would already have a high downward velocity when released and would travel a much shorter distance to the target, making it easier to predict the bomb's path. American dive bomber pilot Harold Buell described the process as "shooting" a bomb at the target[7].

By contrast, the erudite practice of high-level bombing required accurate knowledge of a bomb's physics if its path were to be predicted reliably enough for proper arrival at its destination. Although bomber crews were denied the joy of making the calculations, a bomb's physics had to be precisely designed into the bombsight and the altitude and speed of the aircraft carefully controlled for the bombsight to work.

A bomb dropped from 9,840 feet (3000 m) takes about 25 seconds to reach the ground whether dropped from a moving airplane or a stationary blimp (see Pearl Harbor Bomb Drop Calculations). Combining the bomb's constant horizontal velocity with the ever-increasing downward velocity caused by gravity would make the bomb fall in a downward sloping parabolic path. The situation is similar to drawing with the popular toy called Etch a Sketch®. Turn one knob and a horizontal line appears on the screen. Turn the other knob and a vertical line appears. Obviously, the two knobs are independent. But turn both knobs simultaneously and it's possible to obtain a curved line.

Remarkably, a bomb's motion in the horizontal dimension has no influence over its motion in the vertical dimension. They are like two entirely separate worlds. Speeding up the horizontal velocity will not make the object fall more slowly or more quickly. Likewise, making an object fall in the vertical dimension will not influence velocity in the horizontal dimension.

Ironically, films of real WWII bombing runs, shot through cameras mounted in bomb bays, are often misinterpreted as proof that bombs fall straight down. In the films, the bombs look like they're falling straight down and exploding directly below. But the camera isn't stationary. It's moving forward with the airplane. To give the appearance of falling directly below the camera, the bomb has to be moving forward at roughly the same speed as the camera. The films also confirm that the air resistance slowing the bomb's forward motion is negligible. If air resistance acting on the bomb were significant, the bomb would appear to fall behind the aircraft.

PEARL HARBOR BOMB DROP CALCULATIONS

To calculate the time for a Japanese bomb to fall and strike the USS Arizona, we can use the simple mathematical model, or kinematic equation, shown below. Since this equation will be used in both the vertical (or y-dimension) and the horizontal (or x-dimension), we will use x and y subscripts to denote the respective dimensions.

$$d = \tfrac{1}{2} at^2 + v_o t$$

Where:
d = displacement
a = acceleration
v_o = starting velocity
t = time

Assumptions:
1. The acceleration is constant.
2. There is no air resistance.

In a bomb drop, the starting downward y-dimension velocity is zero. This simplifies the model as follows:

$$d_y = \tfrac{1}{2} a_y t^2$$

To solve for the time the bomb falls before hitting its target, we rearrange the equation and substitute the acceleration due to gravity for ay and the altitude of the bomber for dy as shown:

$$t = (2 \ d_y \ / \ a_y)^{\frac{1}{2}}$$
$$= (2 \ (3{,}000 \ m \ / \ 9.8 \ m/s^2))1/2$$
$$= 24.7 \ sec$$

Now that we know the time, we can switch to the x-dimension and solve for the horizontal distance the bomb travels before hitting its target. In this dimension the bomb starts out moving at 225 miles per hour (101 m/s). Because we ignore air resistance, there is no horizontal force, hence, no horizontal acceleration. (Gravity acts only in the y-dimension.) The equation simplifies to:

$$d_x = v_{ox}t$$
$$= (101 \ m/s) \ (24.7 \ sec)$$
$$= 2{,}490 \ m$$

In the 24.7 seconds it takes the bomb to fall, the bomb travels 2,490 meters, or 1.55 miles. In other words, the bomb has to be dropped 1.55 miles before the aircraft reaches the target in order to hit it.

The logic of using high-level horizontal bombers against ships in the first wave rather than the more accurate dive bombers was simple: the physics were favorable. Having ships tied up at dock simplified the bombing physics, while the physics required to defend the ships was horrific. For starters, shipboard antiaircraft guns were only marginally effective at the distance bombs were dropped. WWII warship antiaircraft guns ranged from rapid-firing .50 caliber machine guns to slow-firing five-inch cannons.

At the moment a bomb was released from a high-level bomber, the aircraft would be 1.55 miles (2,490 m) away, measured horizontally. Antiaircraft guns would have to start shooting long before the airplane reached this point to have any hope of downing the attacker. At such distances only cannons would have had the required range.

Actually hitting a small fast-moving target such as an aircraft at long range is a major physics problem beyond the capabilities of human intuition. An antiaircraft cannon's projectile fired upward would have a noticeable arc caused by the downward force of gravity. The projectile would be less massive and travel much faster than a bomb, making the effects of air resistance significant. To down an aircraft, the projectile and aircraft would have to arrive at exactly the same location at the same time or, at least, come close enough for the projectile to explode near the aircraft.

To make the required calculations, a gunner needed to measure the range, height, and velocity of the aircraft, not to mention have detailed information about the cannon shell's curved path. To make the projectile explode near the aircraft, he had to calculate exactly how long it would take for the projectile to arrive and set the fuse accordingly. He could never aim directly at the target. Instead, he would have to calculate an aiming point that accounted for all the variables.

Making physics calculations using pencil and paper in the heat of battle would, no doubt, have been jolly fun and stress relieving were it not for the time constraints. A Japanese horizontal bomber would have been closing at a speed of 3.75 miles (6.04 km) each minute. If the gunner spotted an incoming bomber at 5 miles (8.05 km) away, he and his crew would have had less than a minute to make the required measurements and calculations, set the fuse, and load and aim the cannon in order to fire before the aircraft dropped its bomb. The results of failing this physics test would be far worse than a failing grade.

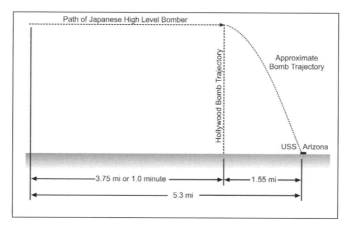

Figure 14: path of bomber

Calculating devices, ironically called "directors," were available during WWII. Directors were mechanical computers that used gears and levers to make physics calculations. Several individuals had to keep the crosshairs of velocity-, altitude-, and range-finding devices aligned with incoming aircraft. These devices fed data to the director, which processed it and provided gun crews much needed information about where to aim and how to set their cannon shell fuses. Unfortunately, the antiaircraft directors weren't much better than Hollywood directors at making accurate physics calculations. A standard U.S. Navy five-inch cannon shell fired at an enemy aircraft had no better than a 0.1 percent chance of downing it during WWII[8].

On the other hand, catching a large stationary battleship by surprise on a clear day was a horizontal bomber's dream. Under normal battle conditions in open water, these ships would be zigzagging while laying down smoke screens to obscure their position and, of course, shooting back in a most uncooperative manner, not to mention having extremely annoying fighter aircraft cover.

In sufficient numbers, American fighter planes could have swept Japanese bombers from the sky, so the first wave of dive bombers focused on destroying the planes before they could take off. The higher accuracy of dive bombers was needed to hit the small targets of airplanes sitting on runways. Because they attack at fairly close range, dive bombers were more susceptible to anti-aircraft fire than high-level horizontal bombers. However, with the element of surprise, a dive bomber could blow up parked aircraft with relative impunity. Compared to ships, air fields were not well guarded by antiaircraft guns. They depended primarily on getting their fighters in the air for protection.

For Americans, December 7, 1941, is an "if-only" kind of date. If only the American aircraft hadn't been parked in easily bombed clusters and had gotten in the air. If only the American military had heeded the signs of an impending attack and stood ready at their guns. If only the DVD had existed, Americans could have corrupted the physics knowledge of the Japanese with Hollywood movies, thereby ruining the aim of their high-level bombers and causing their torpedoes to be harmlessly dropped in the mud.

Such are the thoughts of Monday-morning generals and armchair admirals. Unfortunately, using Hollywood fantasy to counteract physics knowledge is worse than using a knife in a gun fight; it's like using a water balloon in a gun fight. When it comes to real-world tasks, even a small amount of physics knowledge held by a few individuals can overpower a flood of filmmaker foolishness.

BOMB-LIKE JUMPS

Bogus bombing physics isn't limited to just WWII aircraft depictions. From the standpoint of physics, the terrorist who motorcycles off the top of a skyscraper in *True Lies* [PGP-13] (1994) is a falling bomb. In the movie, the motorcycle-riding terrorist (the same bad guy who jumped through the window described in Chapter 5) roars through the lobby of a classy hotel to escape the relentless pursuit of Arnold Schwarzenegger. Admittedly, it's a fantasy far better than running with scissors. Roaring around indoors on a motorcycle vicariously slams all the indoor rules set forth by mommies and grade school teachers everywhere. What satisfaction!

Then Hollywood logic takes over. To escape a pursuer, what should one do? Naturally, go to the highest point in any nearby structure, preferably a perilous place where there is no chance of hiding or escaping. Dutifully, the terrorist rides his cycle into an elevator and pushes the top floor's button.

At the top, he suddenly realizes there's no place to hide or reasonable means of escape. What a surprise. In desperation, he revs up his machine and races over the side. At this point he has a horizontal velocity and a downward acceleration just like the previously described bomb. Like a bomb, his horizontal velocity has no influence over his downward acceleration caused by gravity. By the same token, the downward force of gravity has no influence on his horizontal velocity. Only air resistance can exert a horizontal force. Put the horizontal velocity and ever-increasing downward velocity together and, just like a bomb, the cyclist will travel in a downward-sloping parabolic path.

After going over the side, the bad guy remains airborne for roughly 7.5 seconds. He lands with a slightly downward angle in a swimming pool on top of a shorter building, a considerable distance from where he jumped. The impact does nothing more than get him soaking wet. He walks away without so much as a limp.

A falling motorcyclist is definitely not as aerodynamic as a bomb, but then he is not going as fast. Considering that the terrorist only had about a 66-foot (20 m) runway, his horizontal speed could have been no more than twenty-five miles per hour (40 kph) before his take off, as compared to 225 miles per hour (362 kph) for the previously discussed WWII bomb when first dropped. Poor aerodynamics makes air resistance higher, but lower velocity makes air resistance lower. As a rule of thumb, air resistance goes up by a factor of 4 when velocity is doubled. Low speed would tend to compensate for poor aerodynamics, so it's still possible to evaluate the jump using a simple calculation and ignoring air resistance altogether.

TRUE LIES MOTORCYCLE JUMP CALCULATIONS

We can estimate the height (d_y) of the fall by using the same bomb drop equation derived in the *Pearl Harbor* example:

$$d_y = \tfrac{1}{2}\, a_y t^2$$
$$= \tfrac{1}{2}(9.8 \text{ m/s}^2)\,(7.5)^2$$
$$= 275.6 \text{ m}$$

In other words, the bad guy fell a distance of 904 feet, or roughly 74 stories. His final vertical velocity would have been:

$$v_y = a_y t$$
$$= (9.8 \text{ m/s}^2)\ (7.5)$$
$$= 73.5 \text{ m/s}$$
$$\text{or } 164 \text{ mph}$$

The terrorist's horizontal velocity can be estimated using the distance equation again to estimate his acceleration as follows:

$$d_x = \tfrac{1}{2}\, a_x t^2$$

Rearranging yields:

$$a_x = 2\, d_x / t^2$$

As depicted in the movie, the motorcycle's acceleration on the roof took 4 seconds and occurred in a distance estimated to be 20 meters.

$$a_x = 2\ (20 \text{ m}) / (4 \text{ s})^2$$
$$= 2.5 \text{ m/s}^2$$

The velocity is found as follows:

$$v_x = a_x t$$
$$= (2.5 \text{ m/s}^2)\ (4 \text{ s})$$
$$= 10 \text{ m/s}$$
$$\text{or } 22.4 \text{ mph } (36 \text{ kph})$$

Allowing for some possible inaccuracy in the distance estimate, his horizontal velocity would have been, at most, 25 miles per hour (40 kph or 11.2 m/s). The vertical and horizontal velocities can be combined as follows:

$$V_{TOTAL} = (v_x^2 + v_y^2)^{\frac{1}{2}}$$
$$= ((11.2 \text{ m/s})^2 + (73.5 \text{ m/s})2)^{\frac{1}{2}}$$
$$= 74.3 \text{ m/s}$$
$$\text{or } 166 \text{ mph}$$

Terminal velocity is the highest velocity a falling object reaches and occurs when an object's downward weight force is exactly equal to its upward air-resistance force. Increasing an object's weight or making it more aerodynamic gives an object a higher terminal velocity. Terminal velocity for a human falling with outstretched arms and legs is around 124 miles per hour (200 kph). If the person folds into a ball, the terminal velocity increases to 200 miles per hour (322 kph). Adding the motorcycle's weight to the terrorist would increase the terminal velocity above 124 miles per hour, and so the 166 miles per hour final speed of the cyclist looks reasonable. Using the slowest conceivable velocity of 124 miles per hour, the height would have been overestimated by only 34 percent. In other words, the motorcyclist would still have fallen about forty-nine stories.

The horizontal distance the motorcycle would have traveled, assuming it had an initial horizontal velocity of 25 miles per hour, would be calculated as follows:

$$d_x = v \times t$$
$$= (11.2 \text{ m/s}) (7.5)$$
$$= 84 \text{ m}$$

Based on the analysis and calculations, the terrorist would have fallen seventy-four stories and landed in the swimming pool at about 166 miles per hour (267 kph). When it hit the water, the motorcycle would have slowed abruptly, causing the bad guy's torso to pivot at the waist and violently slam his head forward. He would have ended up wearing a shiny new handlebar mustache in the middle of his face, courtesy of the motorcycle.

It's easy to see that this was not a jump he was likely to walk away from. Furthermore, he would have experienced all the problems a bombardier faces when trying to accurately place a bomb. Even a small error in speed, or aim, could easily have caused him to completely miss the rather small target of a distant swimming pool. To make matters worse, the terrorist had no bombsight. In fact, he would not have even been able to see the pool until he was close to the edge of the building and it was too late to correct his aim.

Aiming issues aside, the motorcycle would probably have fallen far short of the swimming pool. In the 7.5-second jump it would have traveled only 92 yards (84 meters) in the horizontal direction. Judging from the tiny appearance of the people around the pool, the horizontal distance was farther than the 100-yard length of a football field.

If the terrorist actually did fall seventy-four stories, the Marriott Hotel he jumped off would have had to be eighty or more stories high for him to land on top of another building. Keep in mind that the observation deck of the Empire State Building is only eighty-six stories high. Yet, even this height would have been inadequate to give the jumper enough time in the air to travel the horizontal distance needed for reaching the pool.

While *True Lies* does serve up some motorcycle-jumping silliness, at least it does so with a sense of humor. Not to be outdone by a motorcycle-riding terrorist, Arnie, who's riding a horse, attempts the same jump. Unlike the terrorist, the animal has horse sense and stops short, sending Arnie over the side. He's left dangling on the end of the reins. In a parody of old cowboy flicks from the fifties, Arnie finally convinces his not-so-trusty steed to back up and rescue him from certain destruction.

THE POWER OF DIRECTORS

So, why does Hollywood give us these bogus bomb scenes? The answer is a combination of box office savvy and physics ignorance. The director of *Speed* (see Chapter 1) was driving down the highway and saw an overpass bridge with a missing section. Being an imaginative guy in the process of making a movie, he immediately visualized a scene in which his movie's bus would be compelled to jump such a gap. He had no idea if the jump could actually be done, nor did he care. He wanted to do something big for boosting box-office appeal, and his intuition told him this was it. The writers were less than enthusiastic, but that mattered little. On the set the director speaks with the voice of a god.

There were, naturally, problems. As described in Chapter 1, the bridge was flat in the area where the jump was to take place, not to mention that the gap was created on film by carefully erasing the bridge's image. Had there been a real gap, as soon as the bus went off the end it would have, from a physics standpoint, become a falling bomb. The fact that it was moving forward would have in no way stopped or slowed the falling action. As stated in Chapter 1, the bus's wheels would have fallen at least 3.8 feet (1.16 m) below the roadway, causing the bus's front end to slam into the edge when it reached the gap's far side (assuming the bus had reached the unlikely high velocity of 70 mph or 113 kph before arriving at the gap). In the movie, this problem was solved with the distraction of dramatic music, screaming actors, and rapid camera cuts to prevent viewers from focusing on the flatness of the bridge.

Ironically, a real bus jump was also filmed and artfully edited into the movie to give the scene realism. So it's understandable that the reader described in Chapter 1 completely missed the fact that this bus could not possibly have been jumping the gap depicted in the movie. The highly modified bus used in the actual jump drove up a specially made ramp at over sixty miles per hour (97 kph) and traveled over twice as far as the length of the 50-foot gap before slamming into the ground, blowing out its front tires, and destroying its oil pan. Afterward, the bus was undrivable. To the horror of the moviemakers, the bus also traveled so far that it wiped out all but one of the cameras placed in its path. The last camera did get the shot, but it was improperly framed, although eventually used in the movie. All of this could have been prevented if the moviemakers had just made the right calculations. Even more ironically, had the moviemakers spent more time making calculations and studying the physics of the jump, they could have designed a jump in which the bus actually traveled across a 50-foot gap and remained drivable, but that's a subject for the next chapter.

Summary of Movie Physics Rating Rubrics

The following is a summary of the key points discussed in this chapter that affect a movie's physics quality rating. These are ranked according to the seriousness of the problem. Minuses [–] rank from 1 to 3, 3 being the worst. However, when a movie gets something right that sets it apart, it gets the equivalent of a get-out-of-jail-free card. These are ranked with pluses [+] from 1 to 3, 3 being the best.

[–] [–]   Bombs that fall straight down.

[–] [–]   Impossible vehicle jumps.

[–] [–]   Jumps in which a person falls like a bomb for several seconds and walks away uninjured.

[+]   Any of the above that are presented tongue-in-check or with a sense of humor.

CHAPTER 9

Leaping Logic:
Why Moviemakers Say "How High" When the Director Says Jump

JUMPING BUSES

The passengers scream and the driver ducks as the bus hurls towards the edge of disaster. Suddenly, within inches of the gap in the freeway bridge, the front of the bus miraculously flips upward, having hit a short ramp just seconds from oblivion, restoring hope for survival. But when the back wheels approach, the ramp fails. It seems to have disappeared. Instead of being projected upwards like the front, the back wheels go over the edge and fall below it—ending all hope of survival.

The camera angle changes rapidly as the bus drifts across the chasm. When the bus reaches the far side of the gap, are its back wheels even further below the edge? Does it smash into it and explode? Why, no! The back wheels touch down on the roadway. It's a miracle! The columns holding up the bridge have shrunk in height, dropping the roadbed to a lower level.

Yes, the moviemakers did film an actual bus jump of sorts in *Speed* that was then skillfully edited into the film. However, the jump was not made across a gap in a flat section of an overpass bridge as depicted in the movie.

The bus drove at a speed of about sixty miles per hour (97 kph) up a special ramp built on a ground-level section of unused highway. A small additional ramp, called a kicker plate, was positioned at the top and did indeed flip the front of the bus sharply upward as the front wheels drove over it. The kicker then fell out of the way so that it had no effect on the back wheels.

Had the bus merely gone over the ramp, the back wheels would not have immediately fallen below the edge of the ramp. The kicker plate caused the bus to rotate around its center of mass with the front moving higher and the back lower than normal. Since the top of the ramp was about 12 feet (3 m) above the level of impact on the roadbed below, the bus flew horizontally over 100 feet before its rear wheels slammed into the ground, followed by the front wheels slamming downward even more violently. Such a landing left the bus undrivable, a condition that would have doomed the passengers to die in a fiery blast (assuming they were still alive).

To prevent serious injury during the landing, the stunt driver was suspended in a special shock absorbing restraint. Had the stunt driver driven the bus in the normal manner, he would have almost certainly broken his back—an occupational hazard for stunt drivers before the invention of the shock absorbing restraint. As it was, he forgot to wear his mouth guard during the jump and accidentally bit his tongue.

Needless to say, the bus was not an off-the-shelf variety. It was specially modified with driving controls located halfway between the front and back of the bus—a section where normally only passengers sit. This was done to help the driver line up the bus with the ramp as well as put the driver in a less vulnerable position. If the bus went out of control, the front was the most likely part to get smashed in. Everything that could be removed from the bus was taken out to reduce the bus's weight.

SIMPLIFIED BUS JUMP CALCULATIONS

Traditional projectile-motion equations, which ignore air resistance, work well for modeling compact objects such as balls projected off ramps. While ignoring air resistance is not a big source of error for calculating the length of a bus jump, there are other possible errors. When the front wheel goes over the edge, the bus's center of mass is still well behind the edge but is no longer supported by the bus's front wheel. The center of mass essentially has to cross a larger gap than the front wheels. If the bus is not moving fast enough, the bottom of the bus can actually scrape the edge of the ramp.

With the front wheels over the end of the ramp, the back wheels will still be in contact with the ramp and create an upward normal force. This force will tend to rotate the front of the bus downward. On the other hand, the torque applied to the back wheels by the engine will tend to rotate the front of the bus upward. For motorcycle-jump-length calculations, these differences are not a big problem, since the length of the cycle is fairly short. A bus, however, is a lot longer, and making a precise jump length calculation for it would require a computer simulation. Still it's possible, even with a simple equation, to make a reasonable approximation for a bus jump in order to determine if the jump is at all possible.

To start, let's assume that the bus's center of mass is located about half a bus length behind the bus's front tire at the moment the wheel goes beyond the edge of the ramp. When the bus reaches a similar elevation on the other side as it lands, the bus's center of mass will be about half a bus length in front of the gap. So let's model the gap as though it were the length of the actual gap plus the length of the bus. Even when ignoring rotation caused by the normal force on the back wheels and counterrotation from engine torque, this length should yield a conservative estimate of whether or not the jump is possible.

The simple projectile-motion equation for horizontal displacement or range of a jump is as follows:

$$d_x = v^2(2/g)\sin\beta(\cos\beta) \textbf{ (EQUATION 9.1)}$$

Where:

d_x = the range or length of the jump
v = velocity of the bus up the ramp or takeoff velocity
g = the downward acceleration due to gravity, 9.8 m/s^2
β = the ramp angle above the horizontal

Note that the bus's mass does not appear anywhere in the equation. The jump depends only on ramp angle and speed.

Could the jump have been achieved under more realistic conditions, and could the bus have remained drivable? The answer: yes, but with some qualifications. It would have required matching ramps on both sides of the gap and a precise bus speed. Surprisingly, the ramps' angles needed to be no more than 11 degrees and the bus's speed roughly sixty miles per hour (97 kph) in order to make it across the 50-foot gap. Having a landing ramp at the same angle as the takeoff ramp allows the bus to gently touch down, because the bus's net velocity will be nearly parallel to the ramp. This lets the bus roll down the ramp rather than collide with it.

Ramps act like velocity splitters. The takeoff ramp converts part of the bus's horizontal velocity into a vertical velocity component that moves the bus upward and a horizontal component that moves the bus forward. The gravitational force acts in only the vertical dimension and slowly reduces the vertical velocity component

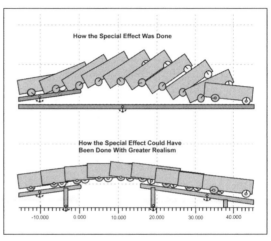

Figure 15: special effect

until it is zero at the top of the trajectory. At the top, the gravitational force then increases the bus's vertical velocity component in the downward direction. The bus has to be up in the air above the takeoff ramp long enough for its horizontal velocity to carry it across the gap.

When the bus lands, it has the same downward vertical velocity as it would have if it were raised using a crane and dropped from the same height as the top of the trajectory. If the bus lands on a horizontal surface, it will slam into it with a considerable force. The bus's downward velocity component will drop almost instantaneously to zero when the bus lands on such a surface, yielding extremely high accelerations and, subsequently, extremely high forces (see Chapter 10).

Here's the big surprise: as long as the bus reaches the correct takeoff speed at the correct takeoff angle (assuming negligible air resistance and vehicle rotation), the bus's mass is not a factor in the length of the jump! Why? It goes back to Galileo, who was perhaps the first person to understand that all objects fall at the same rate regardless of their mass. Aside from takeoff speed and angle, the rate of falling is the primary factor determining the distance of the jump.

Although a jump with a takeoff and landing ramp would have been more realistic than the one filmed for the movie, the double-ramp jump also has its hazards, not to mention problems with reality. If the bus takeoff velocity were too low, the front of the bus would smash into the landing ramp. Too fast and the bus would partially overshoot the ramp and experience a hard landing, be undrivable, and explode or merely break the bus driver's back and seriously injure most of the passengers. Realistically, the speed would have to be higher than the exact level for the crossing because it's better to risk a hard landing than a crash into the landing ramp. Throw in the need to keep the bus drivable and the margin for error is next to nothing.

But all this discussion about adjusting to the precise speed needed for surviving the jump is hypothetical. Takeoff and landing ramps that are exactly the correct angle for making the jump are not likely to be found on overpass bridges. For one thing, changing from an upward to an equal but downward slope in only 50 feet of distance would cause vehicles traveling above sixty miles per hour to go airborne as they crossed the *completed* bridge's peak—a poor design at best. So while the jump may be possible, it's pretty far-fetched.

Still, the true bridge-jump believers are right, in a sense. In theory, the bus jump could have been made even without the ramps—that is, if the bus had been driven fast enough to put it in a circular orbit with a radius equal to the radius of Earth. This jump would have required a takeoff velocity of about 17,700 miles per hour (28,500 kph)—a little quick for most city busses. Air resistance would also have been a factor but might not have been all that bad for a mere 50 feet. The driver and passengers would have likely blacked out from the acceleration required for reaching 17,700 miles per hour, if they survived it. The sonic boom would have smashed windows and rattled nearby buildings but, hey, it would certainly have added excitement. So, in the next movie bus jump, maybe a nuclear rocket scientist will be aboard and just happen to have his latest miniature nuclear rocket creation in his brief case.

JUMPING HULKS

There's no question that the Incredible Hulk is one bad dude and, at first glance, the jumps attributed to him in his movie seem reasonable. But they're not. Such a jump yields projectile motion similar to a bus jump off a ramp. Once airborne the only forces acting on the projectile—in this case the Hulk—are air resistance and gravity, neither of which can help make the jump longer. Sometimes aerodynamically shaped objects such as the discus used in Olympic events can experience lift, a factor that does extend the distance traveled. But lift is unlikely with a boxy object such as the Hulk. This lack of an assisting force after takeoff means that the length of the jump will be dictated entirely by the take off velocity and angle. The movie depicts the Hulk making jumps on the order of a kilometer—an impossibility given his rather slow takeoff velocity.

HOW JUMPING DISTANCE SCALES UP IN CRITTERS

Equation 9.1 established that the length of a jump is only a function of the takeoff velocity squared and the angle. The same relationship is true for cars, critters, or people.

Assume that a critter starts its jump from zero velocity and reaches takeoff velocity by the time its feet (or paws) leave the ground. From kinematics

$$v^2 = 2a(d) \text{ (EQUATION 9.2)}$$

Where:
v = takeoff velocity
a = constant acceleration in the same dimension as v
m = mass of the critter being accelerated
d = distance the leg force acts on the critter before the critter's feet leave the ground

combined with F = ma yields:

$$v^2 = 2(F/m) \text{ d (EQUATION 9.3)}$$

Where:
F = the constant force provided by legs

but

F is proportional to the cross sectional area A of the muscle producing the force. So, again, referring to equation 9.1 for calculating the range or horizontal length of a jump:

The range X is proportional to V^2 or $(A/m)D$
A scales up with the square of the scaling factor (S). m scales up with cube of the scaling factor (S^3) and D with S (see Chapter 4). If an animal is scaled up by the factor S, the new jumping distance will be

$$X_{new} = X_{old} \, (s^2/s^3)s$$

$$= X_{old}$$

In other words, the new jumping distance will be the same as the old, assuming that the animal was not scaled up so much that it collapsed under its own weight.

So, why are tall people often able to jump higher than short ones? They are not scaled up proportionally. Generally, the big difference between short and tall people is the length of their legs. Leg length makes up a larger proportion of a tall person's height as compared to a short person. Proportionately longer legs would be an advantage for jumping because the jumping force they produce when bent and straightened rapidly during a jump would be applied over a longer distance. However, the torso moved by the legs would still weigh about the same as a short person's torso. Although the tall person's legs would weigh more, the torso still accounts for most of a person's weight. Hence a tall person could jump farther than a short one, assuming both had similar athletic conditioning and skill.

So what takeoff velocity would the big guy need to travel 1 kilometer in a single leap, and what would he look like making such a leap? If we ignore air resistance and assume the takeoff angle is forty-five degrees, the Hulk would need a takeoff velocity of about 222 miles per hour (357 kph). He would appear to move away quickly then seem to be slowing down as his image got smaller with increasing distance. While the movie is not a perfect match for the calculated behavior, it's at least in the ballpark, except for one very sticky detail: air resistance cannot realistically be ignored. The Hulk is too large, too boxy, and too fast.

Accounting for air resistance is tricky. First, air resistance changes with velocity—unlike a more cooperative force such as gravity, which remains constant (at least in projectile-motion problems). But it gets worse: at low speeds air resistance can be approximated as follows:

(air resistance) = (coefficient of drag) x (cross-sectional area) x (velocity)

At higher speeds the velocity term in the above equation changes to velocity squared. In other words, the air resistance becomes even more dependent on velocity.

As for the coefficient of drag (CD), it's just a constant, or as engineers call it, a fudge factor—a factor tossed into the equation to fudge the numbers so they come out right. And where would this fancy fudge factor come from? From wind tunnel measurements on the Hulk. That poses a problem: the Hulk is not on hand and probably wouldn't cooperate if he were. To complicate matters further, the CD measurement is only good for one wind direction and one Hulk configuration. When jumping, the Hulk would have to always keep himself oriented in the direction of his velocity and hold his arms and legs in the same position as when his CD was measured. Otherwise, his CD, not to mention his cross-sectional area would change during his jump.

Without turning the analysis into a career, about the best that can be done is to model the Hulk as a sphere and use a computer simulation package such as Interactive Physics to get an idea of how much air resistance affects the Hulk's jumping distance. When we do so, we get an amazing result: the Hulk's takeoff speed must be around 1,250 miles per hour (2,020 kph), faster than the speed of sound, faster even than a speeding .22-caliber long rifle (LR) bullet (873 mph or 1,410 km/hr). Okay, we made a lot of assumptions, but keep in mind that the maximum distance a .22-caliber bullet will travel is only about 1.1 miles (1.8 km). Since the bullet is far more aerodynamic than the Hulk, the Hulk would need to start at a much higher speed to reach even the shorter distance of 1 kilometer (0.62 mi).

So how would the Hulk really look if he made a 1-kilometer jump? He'd look like a giant green cannon ball; he'd go so fast he'd be a blur. His landing velocity (134 mph, or 216 kph) would be far slower than his takeoff velocity due to the effects of air resistance, but when he landed the impact would be impressive.

The Hulk's takeoff would also be dramatic. If he were standing still and decided to leap, he would first bend his knees then very quickly push off. As he straightened to his full height, his feet would break contact with the ground and he'd be launched into the air. The acceleration propelling him into the air would only occur during the short distance between his bent knee and fully straightened position: a distance of, at most, 30 inches (0.8 m). Once his feet broke contact with the ground, he would no longer be able to increase his takeoff velocity. His average acceleration during takeoff would exceed 30,000 gs. The force his feet exerted on the ground would be his normal weight times the acceleration in gs, or about 15,000 tons—enough to break concrete.

Pound for pound, the Hulk's muscles would have to be thousands of times stronger than human muscles to make his lengthy jumps. If an animal is scaled up or down, the distance it can jump (assuming it does not collapse under its own weight) will not change. For example, if a flea can jump 20 inches (0.5 m) in its normal size it will still only be able to jump about 20 inches if scaled up to the size of a cricket. The only way it could jump further would be to get stronger muscles. The Hulk is similar in build to a Neanderthal wrestler on steroids (had there been one). Scale up such a wrestler to the Hulk's size, and he'd still only be able to jump his normal amount—a few meters—not the Hulk's incredible 1,000-meter leap.

So when it comes to jumps in Hollywood movies, the important question is not what the laws of physics say; it's what the director says. When the director says jump, rather than wasting time on calculations, moviemakers simply ask how far and how high.

Summary of Movie Physics Rating Rubrics

The following is a summary of the key points discussed in this chapter that affect a movie's physics quality rating. These are ranked according to the seriousness of the problem. Minuses [–] rank from 1 to 3, 3 being the worst. However, when a movie gets something right that sets it apart, it gets the equivalent of a get-out-of-jail-free card. These are ranked with pluses [+] from 1 to 3, 3 being the best.

[–][–] Creatures making incredibly long jumps at low velocities with little indication of high takeoff or landing force.

[–][–] Hapless souls, heroes or otherwise, cast through windshields for no good reason.

[–] Large creatures with leaping abilities far greater than the smaller versions they were scaled up from.

[–] Simulated vehicle jumps that depart from reality.

[+] Vehicle jumps filmed under realistic conditions.

CHAPTER 10

ACCELERATION AND NEWTON'S SECOND LAW:
How to Get Started, Use the Brakes, or Change Direction, Hollywood Style

NEWTON'S SECOND LAW—A SYNOPSIS

Newton's second law rests on the definition of acceleration, which like most things in physics doesn't have the same meaning as in everyday language. Like force and velocity, acceleration is one of those strange quantities called vectors—represented by arrows. The arrow indicates the quantity's direction, and the length of the arrow indicates the quantity's magnitude or size. Acceleration is simply a measure of how fast the velocity is changing. A change in velocity can mean that an object is slowing down, speeding up, or changing direction. If the change happens quickly, we get a high acceleration; and if it happens slowly, we get a low acceleration.

Newton's second law teaches that

Force = (mass) × (acceleration)
or
$$F = ma \textbf{ (EQUATION 10.1)}$$

In other words, force and acceleration are directly proportional. They go hand in hand. If one increases, the other must also increase; if one decreases, the other must also decrease. The arrow representing acceleration and the arrow representing force always point in the same direction. By contrast, the arrow representing velocity can go in an entirely different direction from the arrows representing the force or acceleration.

If the arrow representing acceleration or force points in the opposite direction from the arrow representing velocity, the object is slowing down. If the two arrows point in the same direction, the object is speeding up. For physics purists, the term deceleration—gasp!—doesn't exist. Okay, some physics books regrettably use this abominable term, but to the pure of heart, it's bad form.

To many, it seems like Newton's second law is just a repeat of Newton's first law in equation form. There's some truth in that, but there's an additional difference: a complete description of Newton's first law defines something called an inertial frame of reference, which must exist for Newton's second law to be true. A frame of reference is whatever is assumed to be stationary (often the floor). An inertial frame of reference is one where Newton's first law holds true.

Is there a place where Newton's first law doesn't hold true? Yes, and it's found even in some fairly ordinary places. An inertial frame of reference can be moving at constant velocity but cannot be accelerating with respect to any other inertial frame of reference. An inertial frame of reference cannot, for example, be a merry-go-round because its parts are changing direction or accelerating with respect to the ground as they go around and around. If a person riding on one side of a merry-go-round tries to throw a ball straight across to a person riding on the other side, the ball will appear to go in a curved path relative to the rotating merry-go-round. Even Earth's surface cannot be considered a true inertial frame of reference because it's rotating. But don't worry; Earth is so large that it can be considered as though it were a flat immobile surface, at least for all the examples in this chapter.

HOW TO ENJOY A CRUISE SHIP CRASH

A runaway cruise ship (*Speed II* [PGP-13])—its engines unstoppable—is headed straight for a dock in the greatest movie ship crash scene ever filmed (not that ship crashes are common). Its passengers scream as the boat rips through the wooden dock. Conveniently, the first mate calls out the boat's speed to reassure viewers that the boat is indeed slowing down (duh). In the middle of the crash, for added excitement (as if there weren't enough), the movie's heroes are hurled through the boat's windshield onto the deck below. The ship's windshield—made of laminated glass to resist wave impact during storms—would lacerate the flesh, shatter the face bones, and knock out the teeth of any unfortunate soul who smashed through it. But miraculously the heroes are uninjured. After taking out the dock and a few condos, the giant boat comes to rest.

So, why were the people aboard the craft screaming? They should have relaxed in a deck chair, sipped a cold drink, and enjoyed the spectacle. There never was any danger, at least not to them. As for the heroes crashing through the ship's windshield, consider the same situation for a car traveling forty-five miles per hour (72 kph) that hits a brick wall and stops almost instantaneously, in say 0.01 seconds. The driver (who considers seat belts unmanly) crashes through the windshield. By contrast, the boat is traveling all of seven miles per hour (11 kph) and takes thirty-three seconds to stop. Common sense alone says that the boat crash is incredibly gentle compared to the car wreck.

A quick calculation shows that the stopping acceleration is over 200 gs for the car and around 0.01 gs for the ship. People will stay put in the car or boat if they have exactly the same acceleration as their respective vehicle. If not, the people will continue moving forward as the car or boat stops. We say that they are "thrown" forward, but really they're not. They passively continue moving forward until a force or combination of forces acts to stop them: in the case of the

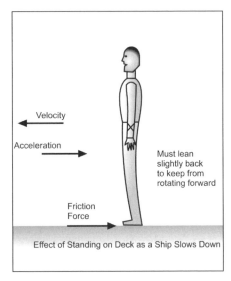

Figure 16: velocity/acceleration man

car's driver, crashing through the windshield and smashing into the wall beyond it provides the combination of restraining forces. In the more gentle case of the ship, the people will never move forward with respect to the ship as long as the friction force between the passengers and deck is high enough to restrain them.

The maximum friction force acting on passengers standing on the deck would typically be at least half their weight. Such a force could keep a passenger fixed to the deck with ship stopping accelerations up to 0.5 gs, or about fifty times greater than depicted in the movie. While it might be difficult to keep one's balance under such conditions, as soon as one fell to the deck, the friction force would be restored and any forward motion with respect to the deck would cease.

The windshield of a ship would be capable of providing a very high stopping force. For the windshield to break from human impact, the ship would need a stopping acceleration of at least five gs. At a speed of seven miles per hour (11 kph), the ship would have to slam into a perfectly solid barrier and come to a complete stop in a remarkably small distance of 3.8 inches (9.6 cm). This distance includes any deformation or crumpling of the ship's front.

Is there any way the giant ship could have momentarily experienced a stopping acceleration high enough to send the heroes through the window? No. A high acceleration requires a high rate of change in motion, but Newton's first and second laws say the ship's huge mass will strongly resist any change in motion. If the velocities shouted by the first mate as the ship slowed down are plotted against time, remarkably, they fall in a straight line. The slope at any point along the line is equal to the ship's acceleration at that moment. But since it's a straight line, the slope is constant, and so is the acceleration. This steady acceleration is way too low to send the heroes through the windshield.

The force that the windshield's glass would have needed to restrain a hero's forward motion would have been equal to his mass times his acceleration. For a 150-pound man during the *Speed II* ship crash, this works out to a whopping 1.5 pounds.

Conversely, Newton's third law says that the man exerts an equal but opposite force on the glass. Surely a windshield could resist a 1.5-pound force. Surely a ship demolishing a dock and several buildings is exciting enough even without propelling the heroes through the ship's windshield.

USING NEWTON'S SECOND LAW

As mentioned in the text, if a car traveling forty-five miles per hour (72 kph, or 20 m/s) hits a brick wall and stops almost instantaneously, in say 0.01 seconds, the acceleration is as follows (assuming acceleration is constant):

$$a = (v_f - v_o) / \Delta t \text{ (EQUATION 10.2)}$$

Where:
a = acceleration
v_o = starting velocity
v_f = final velocity
Δt = time interval
a = (0 m/s—20 m/s) / 0.01 s
   =–2000 m/s^2 or 204 gs
(Note: to obtain gs of acceleration divide 2,000 m/s^2 by 9.8 m/s^2.)

The acceleration just happens to be negative in this case. A negative sign only indicates direction such as forward or backward. Since the velocity was positive, the acceleration had to be negative or point in the opposite direction for the car to slow down. Note also that gs are a unit of acceleration, not force. The force this would create on a 220-pound (100 kg) bubba is as follows:

$$F = ma$$

$$= 100 \text{ kg } (2,000 \text{ m/s}^2)$$
$$= 200,000 \text{ N or } 44,900 \text{ lbs}$$

(Note: as long as the acceleration occurs on Earth, the force acting on a person can be calculated as the person's weight times the number of gs.)

If the initial velocity and stopping distance are known but the time is not, the acceleration can be found as follows (assuming constant acceleration):

$$a = (v_f{}^2 - v_o{}^2) / (2x) \textbf{ (EQUATION 10.3)}$$

where:
x = stopping distance

If a ship slammed into a perfectly solid barrier and came to a complete stop in a distance of roughly 3.8 inches (10 cm) from a speed of seven miles per hour (11 kph, or 3.06 m/s), the ship's acceleration would be as follows:

$$a = (0^2 - 3.06^2 \text{ m/s}) / (2 \bullet 0.10m)$$

$$= -46.8 \text{ m/s}^2 \text{ or } 4.8 \text{ gs}$$

IT'S NOT THE FALL. . . .

It takes no imagination to understand that jumping from a tall building onto a sidewalk is going to do more than just hurt. It turns out that the stopping acceleration is inversely proportional to the stopping distance. When a person hits the sidewalk, it may crack but it is essentially not going to move. The stopping distance has to be entirely provided by crumpling the person. Assume the person is six feet (1.8 m) tall and lands feet first at the terminal speed of a sky diver: 120 miles per hour (193 kph). If the stopping force is provided by crumpling the person 25 percent—in other words crushing the person's legs to about half their normal length—the person will be subjected to 327 gs. At such accelerations, blood flow to the head will stop, blood vessels rupture, and internal organs crash into the bones underneath them[9]. Such a fall is not survivable.

Is any fall from a great height survivable? Based on various unplanned experiments, the answer is yes![10] For example, Lieutenant I. M. Chisov, a Russian airman, was badly injured but survived when he fell nearly 22,000 feet without a parachute after his bomber was attacked by German fighters in 1942. He hit the edge of a snow-covered ravine and rolled to the bottom. In 1944 Nicholas Alkemade—tail gunner in a badly damaged British Lancaster bomber—discovered to his horror that his parachute was in flames after being ordered to bail out. He jumped anyway, landing in trees, brush, and drifted snow. He ended up with a twisted knee and a few cuts. The common element in survival is always an extended stopping distance provided by crumpling materials or objects other than the human who falls.

Water can extend the stopping time, but is of limited benefit. To enter water, a person has to push a volume of water equal to their body's volume out of the way. Since water tends to have a lot of mass, it takes a lot of force to accelerate it out of the way, especially if done quickly. The water exerts a resistance force on whatever does the pushing. Naval academy cadets are forced to jump feet first into water from a ten-meter (32.5 ft) high platform in order to prepare them for the day they might have to abandon ship[11]. They are taught to hop off the platform rather than leaning forward until falling off, because the leaning could misalign them from a perfect vertical position. Any head movement can result in minor injuries such as a bloody lip. Needless to say, above a height of 10 meters, water is going to be a very dangerous landing pad—it only needs to knock a person out to kill them by drowning.

In the movie *XXX State of the Union* [RP] (2005), Darious Stone (Ice Cube) jumps from a train moving at 160 miles per hour across a tall bridge over water. Assuming that the bridge was 300 feet high (91 m) and neglecting air resistance, the hero would hit the water with a vertical velocity of about 94 miles per hour (152 kph). His horizontal velocity would be reduced by air resistance but would probably still be at least as high as his vertical velocity. To have any hope of survival, the hero would have to enter the water feet first to prevent head injury. He'd also have to be facing upward and hit at exactly the same angle as his velocity vector: about forty-five degrees, in order to prevent back and neck injury. Even then, the horizontal component of his legs' velocity would slow down more quickly than the same component of his head and torso, creating a bending action on his body. If facing upward, the body would bend in a direction it's designed for (the same direction as bent in toe touching). Otherwise, the body would be bent backwards, resulting in a back or neck injury.

If the hero entered in the traditional diving position—perpendicular to the water, arms then head—as soon as his arms and head hit the water, their vertical and horizontal motion would

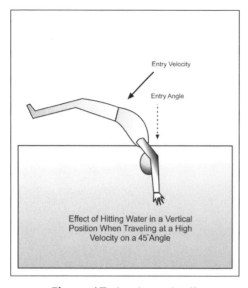

Entry Velocity

Entry Angle

Effect of Hitting Water in a Vertical
Position When Traveling at a High
Velocity on a 45˚Angle

Figure 17: backwards dive

abruptly slow almost to a stop. Meanwhile, his legs and torso would continue with their same vertical and horizontal velocity. The result would be an even more extreme bending action on the body, easily enough to break a neck or even a back and slam the person's torso onto the water's surface, breaking ribs and damaging internal organs in the process.

So what does the hero do? He shoots the water with some type of handgun that looks like a sawed, off grenade launcher and foams up the water below him. Although it's not likely to be effective, foaming the water with gunfire would lower its density and reduce the resistance force it would exert on the hero when he hits it—at least slightly. The hero dives into the foamy water vertically head first. Does he die? Does he suffer? Why no, it's a miracle! He not only remains conscious but survives unscathed. After all, it's not the fall; it's the stop at the end of the fall that kills and, of course, this stop was perfectly safe by Hollywood standards.

DEATH BY RESCUE

Falling off a tall building is almost certain death unless one is miraculously rescued. So, when Lois Lane falls from the fiftieth floor and is inches from impact with the sidewalk, Superman must rush—faster than a speeding bullet—to save her by whisking her off in a horizontal direction. As she falls she will roughly reach the terminal speed of a sky diver and be closing with the sidewalk at 120 miles per hour (193 kph). On the other hand, the man of steel will be closing with her at a velocity in excess of a speeding bullet say around 1,400 miles per hour (2250 kph). When he catches Lois, he must increase her velocity from zero in the horizontal direction to match his horizontal velocity and stop her downward velocity almost as fast as if she had hit the sidewalk. If it takes 0.1 seconds to do this, Lois will be subjected to over 6,000 gs of horizontal acceleration, and Superman will end up with an armful of bloody mush. It makes no difference whether high acceleration occurs in the horizontal or vertical direction. It's going to hurt.

Superman could stop—he is after all superhuman—the instant before he hits Lois, catch her, and then accelerate off in a horizontal direction at a rate that would not injure her. Just before stopping, he would have 10,000 times more kinetic energy than a 7.62 NATO machine gun bullet. The law of conservation of energy demands that something be done with the energy. About the only option is converting it to heat. When he stopped, Superman would become red hot and likely set Lois on fire—not in the romantic sense. Lois would be french fried. And since she's inches from the sidewalk, Superman is still going to have to subject her to a high vertical acceleration to get her stopped—but certainly not 600 gs. There's really no way he can save her, unless he stays close at hand so that he would not have to move so fast to catch her.

Acceleration Injuries

Human tolerance for acceleration depends on many factors, including age, physical fitness, direction of acceleration, and use of safety equipment. The following data is offered only as a rough indication[12].

Blackout from prolonged exposure	4–6 gs
Chest acceleration limit during car crash at thirty miles per hour (48 kph) with airbag	60 gs
Head acceleration experienced by Princess Diana during fatal car wreck	100 gs
Chest acceleration experienced by Princess Diana during fatal car wreck	70 gs

CORNERING CALLING FOR A CORONER

The spacecraft from Earth (SFE) changes its direction to fight an enemy ship. It is traveling a mere 0.25c (25 percent of the speed of light) and makes a gentle 180-degree turn with a 1.0-mile radius (1.6 km). The enemy ship departs slowly without bothering to fire. Has it given up, surrendered, or retreated in fear? No, there is no need for any of these responses. The crew members of the SFE have turned themselves to bloody mush by making the 180-degree turn. They have subjected themselves to roughly 3.6×10^{11} gs of acceleration. Even making the turn at a mere 1,000 miles per hour (1,600 kph) would subject the crew to 12.7 gs of acceleration—enough to cause blackouts and even fatalities. The truth is that space battles would have to be fought at rather sedate speeds if the ships were supposed to make turns and keep the crew alive at the same time.

If a spaceship makes a turn, it has obviously accelerated because it has changed direction. In making the turn the ship will generally follow a circular arc. Anytime an object goes around in a circle, or for that matter even a part of a circle, it will be subjected to centripetal acceleration. And acceleration is acceleration, whether it is caused by traveling in a circle or some other type of activity. High acceleration leads to high forces that cause damage.

If centripetal acceleration is such a big problem for humans on spaceships, then how do air force jets flying at supersonic speeds engage in dogfights? Simple, they don't. First, if they want to down an enemy aircraft, they will typically use a small guided missile, which travels much faster than the jet, locks onto the target, and destroys it when it's several miles away. Small missiles can handle far more acceleration than humans and don't blackout. If a jet fighter does engage in a conventional dogfight with blazing cannons, it has to drop to subsonic speeds (below 760 mph, or 1,200 km/s).

Centripetal acceleration has been a boogeyman for military pilots since WWII. Dive bombing took a special kind of courage. Not only did the pilot have to dive straight at a target that was likely shooting back, but once he'd released his bomb he had to pull up sharply to avoid slamming his aircraft into the target. This subjected him to multiple gs of centripetal acceleration at a most inconvenient time for a blackout. Blackouts were so common among dive bombers that the German Stuka bomber was designed with a type of autopilot that would pull the aircraft out of its dive as soon as the bomb was released.

For spacecraft, even slowing down would be a major problem. If a spacecraft traveling at 0.25c (25 percent of the speed of light) decided to stop and limited its acceleration to 1.0 gs, it would take 3 months to come to a stop. Even if spacecraft could attain incredible speeds, simply stopping would make space journeys lengthy.

If the spacecraft were rotating in order to create artificial gravity (see Chapter 15) and wanted to slow down, it would have to turn off the rotation and reorient all the spacecraft floors so that the downward direction in the spacecraft's rooms pointed in the same direction as the forward motion of the craft. While not impossible, it would take a sophisticated design to pull this off. Stop in thirty seconds from a speed of 0.25c, and the crew would be bloody mush (acceleration = 2.6×10^5 gs), that is, if the spacecraft did not incinerate itself. Its kinetic energy would have to be converted into another form—most likely heat.

Star Trek dealt with this annoying little problem by creating "inertial dampers." These fictitious devices operate on an unknown principle of physics and somehow dramatically reduce the effects of high acceleration. As stated earlier, Newton's second law says that F = ma, and it's actually the F (or force) that causes the damage. The m is generally referred to as the object's mass, but really is its linear inertia or resistance to change in linear motion. Einstein theorized that m could be increased dramatically without adding a single molecule to it. All the object had to do was go fast enough to approach the speed of light and—presto—the object's linear inertia would approach infinity. Surely, if inertia can be increased, then it can be decreased, hence, the creation of a fictitious inertia-decreasing device or inertial damper—a device that stretches physics like a rubber band. Decrease a person's m to nearly zero, and people can tolerate high accelerations because the high accelerations will produce low levels of force on them.

Having the inertial dampers go offline due to damage by attacking enemy warships is standard practice in *Star Trek* episodes. When this happens, the starship's interior jiggles about and crew members fall down. It's all great fun. If such an event actually happened, the crew members would be puréed the first time the spaceship made a high-speed maneuver.

Star Trek is not consistent in its use of inertial dampers, and there is no real scientific basis for them. But the writers have avoided trying to explain them with useless scientific babble. The mere mention of inertial dampers makes it clear that the writers know there are serious problems with high-speed maneuvering. Without the dampers, storyline fallacies would abound. Inertial dampers border on comic-book science, but at times science fiction has to resort to pure fiction in order to have a story.

Summary of Movie Physics Rating Rubrics

The following is a summary of the key points discussed in this chapter that affect a movie's physics quality rating. These are ranked according to the seriousness of the problem. Minuses [–] rank from 1 to 3, 3 being the worst. However, when a movie gets something right that sets it apart, it gets the equivalent of a get-out-of-jail-free card. These are ranked with pluses [+] from 1 to 3, 3 being the best.

[–] [–] Movie characters subjected to huge accelerations with no significant injury.

[–] [–] Movie characters riding in vehicles are thrown through windshields as the vehicle slows down with low-to-moderate acceleration.

[–] [–] Space battles with ultra-high-speed turns and maneuvers and no physiological effects on crew members.

[+] Sparing use of make-believe devices that are not explained with a lot of scientific mumbo-jumbo but at least recognize that a scene's physics would otherwise be impossible.

CHAPTER 11

HIGH-ENERGY FILMS:
Nuclear Firecrackers, Falling People, and Cars as Weapons

NUCLEAR FIRECRACKERS

Up close a nuclear bomb blast is deadly but merciful—instantly vaporizing breathing, feeling humans into shadows on the ground. At a distance, it's pure cruelty. It can melt flesh off a face or cause prolonged torment through all kinds of trauma, not to mention radiation sickness, cancer, or genetic mutation. Few events can match its horrifying effects on people. It's our ultimate boogeyman, yet our ultimate bodyguard. We can choose where, when, and how to unleash its incredible power. And when we do, anything or anybody that threatens us had better watch out.

So, of course, when giant spaceships, enormous asteroids, or huge masses of uncooperative molten iron threaten Earth's tranquility, it's time to reach for the nuclear button. The problem is that on a planetary scale, our boogeyman bodyguard is a squeaky little mouse.

Oversized saucers 15 miles in diameter, sent by a mother ship, arrive from outer space in fiery clouds and park over major cities around the globe. At first, the inhabitants below the saucers respond with a mix of excitement, curiosity, and fear. But when the saucers start blowing up cities with single blasts from their death rays (*Independence Day* [RP]) the confusion vanishes—along with the White House. It's show time for the nuclear bomb, our boogeyman bodyguard. But wait! David Levinson (Jeff Goldblum), always the devoted environmentalist, shouts "stop." Nuking alien invaders over American cities would be an environmental disaster. How could anyone unleash such a boogeyman? On the other hand, letting alien ships incinerate entire cities complete with birds, trees, and nuclear power plants is—by golly—also an environmental disaster. Less emotional military officials (nitwits according to Hollywood) decide to nuke the aliens anyway, to no useful effect. It seems the saucers have protective force fields. As Goldblum predicted, Earth gets the boogeyman but not the bodyguard.

What to do? For every Goliath there's a stone. An alien saucer must turn off the shield around its death ray just before firing it. Fly a jet aircraft up the death-ray port and voila: the alien ship explodes and falls down, all 15 miles in diameter of it. Krauss, in *Beyond Star Trek*, estimates a saucer would weigh about 100 billion tons and that dropping it from a height of about 1 mile would release over 10,000 times as much energy as the nuclear bomb used on Hiroshima.

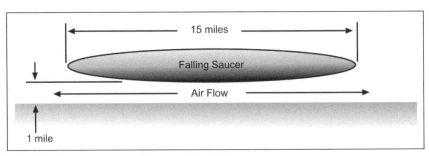

Figure 18: flying saucer

It's hard to imagine that an object falling from a height of 1 mile would create the equivalent of a gigantic nuclear bomb blast—but then it wouldn't. The blast would be worse than an exploding bomb. A falling 15-mile-diameter disk would act like a gargantuan piston. Air underneath would be forced sideways out of the gap between the piston and the ground, forming a horizontal blast wave in the process. A nuclear bomb releases energy upward as well as horizontally, but it's the horizontal energy that destroys cities and countryside. The falling saucer would waste little of its destructive energy in the vertical direction. The drop would mostly unleash energy in a horizontal blast wave.

The air velocity out the gap would be equal to the downward velocity of the saucer times the ratio of the saucer's area to the area of the gap at the saucer's perimeter (assuming that the air underneath acted as though it were incompressible). As the saucer fell, the gap would get smaller and smaller. When the saucer was 0.25 miles above the ground, the area of the saucer would be fifteen times as large as the area of the gap. If the saucer were falling at about half the terminal velocity of a sky diver, sixty miles per hour (97 kph), the horizontal wind coming out of the gap would be 900 miles per hour (1450 kph). These velocities are easily comparable to those of the Hiroshima nuclear blast[13]. True, air is compressible and so the wind will be lower, but the gravitational potential energy of the spaceship that is not converted into wind speed will be converted into elevated pressure and temperature beneath the spaceship, both of which cause damage.

Make the same calculation at a height of 100 feet, and the area of the saucer would be 198 times as large as the area of the gap. The wind velocity would be nearly 12,000 miles per hour

(19,300 kph)—an impossibly high number. Obviously, instead of the extreme wind speeds, air would now be compressed to very high pressures, attaining high temperatures in the process. Combustibles under the saucer would likely ignite and blast a high-speed wall of fire out the sides of the gap. Unlike a nuclear blast radiating in all directions from a point source, the saucer blast wave would occur along its 47.1-mile perimeter and be focused in a horizontal direction. The effect would devastate a much wider area than an equivalent nuclear bomb blast.

Okay, the saucer would probably tilt as it fell, but if it's only 1 mile off the ground, the tilt would be less than seven degrees. If it fell from a height of, say, 15 miles, the tilt could approach ninety degrees, but then the ship would also have fifteen times as much potential energy. Besides, if the ship tilted as it fell, the first side to touch the ground would block air flow, meaning that airflow out the elevated side would be higher than if the ship did not tilt. Tilting is going to do little or nothing to moderate the severity of the disaster.

To be capable of traveling back to its mother ship, the saucer would need fuel with at least 1,000 times as much energy as the blast it set off by falling 1 mile. Imagine the environmental damage this would cause when it leaked out and/or exploded during the ship's crash. Killing these Goliaths all around the globe would be no cause for celebration. It would change weather patterns, ignite massive fires, and fill the sky with debris that would blot out the sunlight.

As for the mother ship, while the Davids on Earth were readying themselves to fight the Goliath saucers, the nerdy Goldblum and ace pilot, Captain Steven Hiller (Will Smith) flew a captured alien fighter craft to the mother ship, bluffed their way inside, planted a virus in the mother ship's computer, and left a—you guessed it—nuclear boogeyman. The pair made their escape just before the entire mother ship detonated. Talk about a lucky rock to the temple. The ship is one-fourth the size of the moon with an outside surface area nearly four times the land area of Texas. Under normal circumstances, blasting such a colossus with a single nuclear warhead would be as effective as tossing a firecracker at a hornet's nest—just enough damage to make them really mad. But here the boogeyman apparently sets off the alien ship's fuel supply, destroying the entire mother ship in the process.

The explosiveness of the mother ship's fuel does not bode well for resistance forces back on Earth. Assuming the saucers' fuel is the same as the mother ship's, when the saucers crashed, their fuel would most likely explode, making a bad situation worse. Humanity would be lucky to survive. But there's no such unhappy ending in the movie. The heroes return victorious, minus the usual brave but minor characters sacrificed to provide just the right touch of sadness, and humanity is saved.

INCOMING ASTEROIDS

In *Armageddon* [RP] a Texas-sized asteroid—bigger than Ceres, the largest known asteroid in the solar system—is headed toward Earth at a speed of 22,000 miles per hour. The usual lovable assortment of misfits and neurotics in a complete spectrum of shapes and sizes is gathered, trained, strapped into space shuttles and sent to drill an 800-foot-deep hole, plant a nuclear bomb on a convenient fault line, and split the asteroid in half—the preparations having been accomplished in a mere eighteen days. Curiously, this gigantic asteroid has been able to sneak into the solar system on a collision course with Earth, travel for around two decades, and avoid detection until it's almost too late.

First, imagine an 800-foot hole (hardly Texas-sized) compared to the size of the asteroid it's drilled in. Then imagine the overall damage if a nuclear bomb went off in such a hole drilled in "nowhere" Texas. The result: not much. Okay, maybe there would be some radioactive fallout and a big blast hole, but the Lone Star State is still going to be largely intact. It doesn't seem likely that the same setup in a Texas-sized asteroid is going to do much more damage.

Of course, the movie's military morons—there hasn't been an intelligent movie general since *Patton* [PGP] (1970)—fail to appreciate the importance of the last few feet of depth in the hole. In fact, they fail to appreciate the hole. When the project falls behind and the last possible moment for detonation is fast approaching, they try to explode the bomb even though it has not been properly planted.

An asteroid the diameter of Texas (diameter = 1,271,900 m) with the same density as Earth would have a mass of about 6×10^{21} kilograms. The largest nuclear bomb ever built was a Russian device rated at 100 megatons TNT and weighed a whopping 60,000 pounds (27,000 kg), nearly the 24,400 kilogram payload of a space shuttle. If the bomb were miniaturized and used on the asteroid, and all of its energy went into pushing the halves apart with no energy wasted on the splitting, each half would end up with a velocity of 0.02 miles per hour (0.03 kph).

The drillers land on the asteroid after it has traveled past the moon, giving them around ten hours before it hits Earth. Allowing eight hours for drilling leaves only two hours before the halves reach Earth after the bomb is set off. Multiplying this time by the separation velocity of the halves equals the separation distance when they reach Earth: a whopping 66 yards (60 m) apart (assuming no gravitational attraction force between them).

A computer simulation is needed to account for gravitational attraction between the halves. Such a simulation calculates that the asteroid halves require a separation velocity of 4,738 miles per hour (2,119 m/s) to miss Earth by the 400 miles stated in the movie. This means that each asteroid half would have to gain 6.7 $\times$ 10^{27} joules of kinetic energy, requiring at least 64-billion 100-megaton nuclear bombs to do so. Earth's gravitational pull would cause the asteroid halves to slingshot around the Earth and collide back together on the other side. The kinetic energy gained to separate the asteroid halves would then be converted back into heat about 1,000 miles above the Earth's surface. The release of energy would be astonishing and likely cause fires and damage on Earth directly beneath the blast.

Here's the scary part: slapping together a plan for destroying even a small incoming asteroid in a few weeks time and splitting or vaporizing it with a nuclear bomb is ridiculous. A realistic movie portrayal of Earth's current ability to deal with asteroid strikes would have been a public service and a much needed wake-up call. Such a movie could not come from Hollywood, however, because it would lack the required happy ending. *Armageddon* makes no such errors. Its plucky drilling crew overcomes all the obstacles and blows the asteroid in half, albeit at the cost of several crew members. Nevertheless, this loss adds just the right amount of pathos to the joyous ending as humanity is saved, yet again.

RIDICULOUS ROTATION

Not to be outdone by other nuclear bomb follies, *The Core* [XP] has the Earth's core stop rotating—caused, naturally, by military morons playing with their new earthquake weapon, which was invented by a brilliant scientist with an ego as big as a house and no head for long-term consequences. In this movie it takes a mere three months to build a magical ship called the Virgil that can carry a rescue team down to the core for restarting it with, guess what, nuclear bombs.

At least this time they send five bombs. If we assume that the bombs are 100 megatons each and 100 percent of their energy goes into restarting the core, the crew is still short by at least 685 bombs. Keep in mind also that currently the biggest bomb in the U.S. arsenal is only nine megatons.

There are other small details such as creating the twisting action, or torque, required to get the core spinning. Exploding nuclear bombs tend to create force in all directions, which would only produce a net force acting directly through the core's center of mass. Such a force cannot cause rotation. For rotation, the force must be applied at a ninety degree angle to the core's radius in order to produce the needed torque.

As usual, nothing goes as planned and the mission reaches the brink of failure. The movie's five-star moron decides to restart the core by using the earthquake machine in reverse. This seems like a much better plan than drilling down to the core, but, alas, an enlightened *Virgil* crew member warns that it will cause all of Earth's volcanoes to erupt. Use of the earthquake device would also doom the *Virgil* and its remaining crew members. Fortunately, the military moron's scheme is foiled and the crew triumphs again, at a cost. The older, less good-looking crew members die, once more providing just the right mix of pathos and triumph for the joyous ending.

All of this assumes that Earth's core could stop rotating while the crust and mantel continued happily spinning—an assumption that goes beyond silliness. There would have to be a zero viscosity, friction-free layer between the mantel and core for this to happen. The rotational kinetic energy of the previously rotating core would also have to go somewhere. About the only choice would be to turn it into heat. Surely an extra 69,000 megatons of TNT worth of heat appearing inside the Earth would cause earthquakes, volcanoes, tsunamis, or something. Certainly, it would take more than a military gismo to cause it and more to fix it than a device that, on a global scale, amounts to a military firecracker.

FALLING HUMANS

It's a deadly but simple principle: the gravitational potential energy stored in an object becomes kinetic energy during a fall. It's the same type of energy that makes bullets lethal.

Kinetic energy is calculated as follows:

$$(kinetic\ energy) = \tfrac{1}{2}\ (mass)(velocity)^2$$

A .45-caliber bullet, for instance, has a mass of 0.015 kilograms and a muzzle velocity of around 254 meters per second, giving it a kinetic energy of 483 joules (Remington Express 230 MC). For comparison, let's assume an action hero with a mass of 49.3 kilograms (108 lbs) falls out of bed. The bed is an old-style, double-post model that Lincoln could have slept in—a little higher than normal—say one meter high. Her gravitational potential energy in bed can be calculated from the following:

(Potential Energy) = (mass)(g)(height)

The gravity constant g is 9.8 meters per second squared (using metric units) on Earth. Thus, her gravitational potential energy in bed is 483 joules. Since this energy is converted into kinetic energy during the fall, the hero hits the ground with the kinetic energy of a .45-caliber bullet!

The hero lives because the energy of the fall is dissipated over a much larger area than the area of a bullet. But if you ask around, it's usually easy to find a friend or acquaintance who has suffered a broken bone from a fall of a similar height. Elderly people, in particular, are vulnerable to such falls.

Each additional meter of height is like adding the kinetic energy of another .45-caliber bullet. From a kinetic energy standpoint, a mere 6-meter (19.8 ft) fall—routine for an action hero—is similar to being simultaneously shot by six .45-caliber bullets.

Suppose our hero is a 109-kilogram (240 lb) body builder instead of the wiry person mentioned in the first example. Now a 6-meter fall is like getting shot simultaneously by thirteen .45-caliber bullets at point-blank range. Indeed, it's true that the bigger they are the harder they fall.

Yes, bullets are incredibly lethal because they can penetrate into vital organs and a fall on a sidewalk would lack this penetration. But, it's pretty hard to completely avoid injury from being simultaneously shot point blank six times with a .45-caliber let alone thirteen times, even when wearing a bulletproof vest. A 6-meter (19.8 ft) fall directly onto a sidewalk would almost certainly break bones even with a good landing. Increase the height beyond 6 meters, and it could easily be fatal.

FALLING BULLETS

While humans falling from great heights are almost certain to be killed, humans struck by bullets falling from great heights can survive. To help understand the reasons, we can compare the terminal velocity of a falling bullet with that of a falling human. A human skydiver will reach a terminal velocity of around 120 miles per hour (193 kph). At this point the upward air resistance force is equal in size to the downward gravitational force on the person, giving a net force of zero; hence, the person cannot accelerate to a faster speed. Traveling at terminal velocity, a 154-pound (70 kg) person has a kinetic energy of 209 .45-caliber bullets. He will literally explode when hitting an unyielding surface such as a sidewalk.

By comparison, when a .45-caliber bullet is fired directly upward, it will go up at a very high velocity but come down at a terminal velocity only slightly higher than the sky diver—about 170 miles per hour (242 kph). At this speed the bullet will have a kinetic energy of only 43 joules, about 9 percent of its original kinetic energy and less than the average energy required to produce a disabling wound, 81 joules (60 ft-lb), as reported by U.S. Army ordnance expert Major General Julian S. Hatcher (Hatcher's

Notebook, 1962). However, the term average implies that sometimes a lower value is sufficient to produce a disabling wound. Also chances are very high that the bullet will strike its victim in the head, and no blow to the head can be considered harmless.

Indeed, doctors at the King-Drew Medical Center in Los Angeles claim to have treated 118 people for falling-bullet injuries (including 38 fatalities) between the years 1985 and 1992, mostly attributed to the massive discharge of firearms during celebrations such as New Year's Eve[14]. These statistics suggest that about a third of the people hit by falling bullets die, but many questions have been raised about the validity of the numbers. First, minor injuries are typically not reported. Second, the criterion for listing a random falling bullet as a cause of a gunshot wound is very liberal, and witnesses are often less than forthcoming with details.

There are also many factors that can increase the harmfulness of a random falling bullet. Common rifle bullets, for example, tend to be more aerodynamic and come down at higher velocities than large-diameter bullets such as those of a .45-caliber handgun. When they strike a victim, longer and thinner rifle bullets dissipate more energy per unit of area than shorter and fatter handgun bullets, and are more likely to penetrate the skin. But the single biggest increase in danger is the fact that drunken revelers firing bullets in the air are not noted for their careful aim. They often don't fire their bullets straight up into the air. Fired at even a slight angle, projectiles will come down nose first instead of base first, decreasing air resistance and increasing bullet speed in the process. If the angle is significant, the bullet can come down at a substantially higher speed.

The Mexican [PGP] is about the only movie to make a plot device out of a fatality caused by a falling bullet from a horde of drunken gun shooters. Okay, it's certainly possible but not predictable. On the other hand, imagine the plot possibilities for falling bullets on planets or moons with no atmospheres.

An assassin pauses next to the rover parked on the moon's surface and carefully places a box beside it. He presses a button and steps back as the box silently emits puffs of smoke from its top for about two seconds. Concealed inside is a Mac 10 submachine gun with a special aiming mechanism that keeps it pointed exactly upward, and a solenoid to depress its trigger. The assassin removes the box and disappears. About 5.4 minutes later the victim starts working on the rover as thirty .45-caliber bullets rain down on his head. Since there is no air resistance, they strike the victim's space suit at the same velocity they left the barrel of the Mac 10 a few minutes earlier. They hit with a slight amount of scatter caused by the recoil-induced vibration of the submachine gun, but there is no wind to blow them off course. The lack of an atmosphere opens up all kinds of creative possibilities for plot devices.

CARS AS WEAPONS

The hero walks down the narrow alley and looks behind as 3,000 pounds (1,360 kg) of dark sedan roars toward him at thirty miles per hour (48 kph) from a distance of 100 feet. At the last instant, he jumps out of the way—or, at the last instant he is hit and rolls over the top of the sedan. In either case, he grimaces, dusts himself off, and goes on his way. He is after all the hero. One wonders why the bad guys keep trying assassination by automobile. It never works.

Although the hero had a mere 2.3 seconds to assess the situation and make his move, it's no problem. He's handled worst. But the car's kinetic energy is not so easily dismissed. It is the equivalent of 254 .45-caliber bullets. This is more kinetic energy than a 154-pound (70 kg) person falling at terminal velocity (209 .45-caliber bullets). The stop isn't as sudden as hitting the sidewalk, but rolling over the top of a car is no roll in the hay. It's going to do some damage, and portraying it otherwise is going to do some damage to clear thinking.

In the real world a drunken imbecile weaving his car down the road at 30 miles per hour should be greeted with even more horror than a drunken fool emptying his six-shooter skyward in the middle of the street. Maybe it's not Hollywood's place to help us put the dangers in perspective, but it certainly wouldn't hurt clear thinking if car-pedestrian interactions in movies were more injurious.

Summary of Movie Physics Rating Rubrics

The following is a summary of the key points discussed in this chapter that affect a movie's physics quality rating. These are ranked according to the seriousness of the problem. Minuses [–] rank from 1 to 3, 3 being the worst. However, when a movie gets something right that sets it apart, it gets the equivalent of a get-out-of-jail-free card. These are ranked with pluses [+] from 1 to 3, 3 being the best.

[–] [–] [–] Using underpowered nuclear bombs to save humanity from certain destruction.

[–] [–] [–] Contrived happy endings that have no logical or reasonable scientific basis and that help obscure real problems facing humanity.

[–] [–] Having massive objects fall to Earth with only minor consequences.

[–] [–] Slapping together major engineering projects in ridiculously short periods of time.

[–] Humans falling from great heights and receiving no injuries.

[–] Moderate- to high-speed car-pedestrian accidents without so much as a bruise on the pedestrian.

[0] Fatalities from falling bullets on Earth (possible but unlikely).

[+] Fatalities from falling bullets on moons or planets with no atmospheres.

CHAPTER 12

MOVIE MOMENTUM:
The Attractive Force of Glass, Rail-Gun Recoil, and Cosmic Toyotas

SHOTGUN BLASTS AND THE ATTRACTIVE FORCE OF GLASS

Sergeant Martin Riggs (Mel Gibson) stands on the sidewalk as a sinister car approaches with a shotgun protruding from the window. Suddenly he sees it, but—blam!—too late. He's blown violently off his feet and flies several feet backward through the nearest display window. Fortunately, he's wearing his bulletproof vest and survives (*Lethal Weapon* [PGP-13] 1987).

If he were not on the sidewalk by a display window, then invariably he'd be blown into a rack of whisky bottles, a giant mirror, or some other large glass object. This happens so often in movies that Hollywood seems to have discovered a new principle of physics: the attractive force of glass. Fortunately, under normal circumstances, one needn't fear. Although remarkably reliable, the attractive force of glass only works for shooting victims and then only in movies. Still, wouldn't something as deadly as a shotgun blast at least blow a victim violently backward?

At first glance it looks like it would, and that we could prove it using the law of conservation of energy. We could calculate the buckshot's kinetic energy before colliding with the victim and confidently predict that the victim would end up with this energy after being shot. If the energy were still in the form of kinetic energy, the victim's velocity could then easily be calculated. Indeed, in an elastic collision this analysis would work perfectly. The trouble is that in an elastic collision the objects that collide don't stick together.

A load of buckshot hitting a vest tends to stick; hence, its collision with the victim is clearly inelastic, meaning that the kinetic energy of the victim will always be significantly less than the original kinetic energy of the buckshot. The "lost" kinetic energy is not really lost but rather altered in form. Some energy becomes a shock wave in the victim, creating bruises and possibly cracked ribs. Some energy converts immediately into heat. Predicting where all the kinetic energy goes and what it does is a daunting task, and so a kinetic-energy analysis can't determine whether a victim will be blown backward.

A quantity called momentum, however, is surprisingly helpful in these calculations because it cannot change form. Momentum is a measure of how hard it is to stop an object. Conveniently, when an object, such as a bullet, collides with another object, such as a shooting victim, the bullet shares its momentum with the victim. In other words, the bullet's momentum immediately before the bullet strikes must be equal to the momentum of the bullet and victim immediately after, regardless of what happens to the kinetic energy. The phenomenon is called the law of conservation of momentum and as noted in earlier chapters, any law with the word conservation in it is about as close to absolute truth as we humans can get.

Unfortunately, friction messes up conservation of momentum. So, how do we handle this pesky little force? We simply ignore it. Okay, it sounds pretty sloppy, but in reality it isn't. First, if the analysis is made immediately after the collision, friction will not have had enough time to mess things up significantly. Second, if the victim is blown off his feet, there will be little friction force present while he's flying through the air. Third, if a friction-free calculation says a victim will not be blown violently backward, then accounting for friction will make the event even less likely: the friction force would resist backward motion.

ANALYSIS OF THE BACKWARD MOTION OF A SHOOTING VICTIM

We'll assume there's no friction to impede the backward motion of the victim. This would favor the event's occurrence. To calculate the momentum of an object, we use the following equation:

$$p = mv \text{ (EQUATION 12.1)}$$

Where:
p = momentum
m = mass
v = velocity

Before the buckshot collides with the victim, the victim's momentum is zero, since he's not moving. This means that we only have to consider the forward momentum of the buckshot. For simplicity we'll treat the buckshot as though it's a single object rather than calculating individual momentums for each pellet and adding them together. Both methods give the same result.

After the collision, the victim and buckshot stick together and so, again, we only have to calculate the momentum of their combined mass. We'll assume that the combined mass of buckshot and detective is $M_D = 80$ kilograms, and the momentum after the collision as P_D even though it's really the momentum of both the detective and buckshot stuck together. From conservation of momentum:

$$\Sigma p_{BEFORE} = \Sigma p_{AFTER}$$

or

$$p_b = p_d$$

Substitution yields:

$$m_b v_b = m_d v_d$$

Note that the velocity of the detective is proportional to the ratio of the buckshot's mass to the detective's mass. This ratio is going to be tiny.

$$v_d = m_b / m_d \bullet v_b$$
$$= (0.0318 \text{ kg}) / (80 \text{ kg}) \bullet (486 \text{ m/s})$$
$$= 0.193 \text{ m/s, or about 0.4 mph, a pretty lame speed}$$

If we run a momentum analysis (see "Analysis of the Backward Motion of a Shooting Victim") on the hapless Sergeant Riggs, we find he's blasted backward at the momentous speed of 0.4 miles per hour. Keep in mind that humans can walk briskly at about 4 miles per hour. Since the analysis was done with assumptions that favor being blown backward, it's clear that not just Sergeant Riggs but shooting victims in general aren't going to be blown backward by the force of a shotgun blast.

Here's another way to analyze the situation: apply the law of conservation of momentum to the shooter similar to the way it was applied to the victim. In other words, recoil from firing a weapon will give a shooter backward momentum equal to the forward momentum of the buckshot and hot gasses (from burning gun powder) exiting the shotgun's barrel. (Note: buckshot will also include a light-weight, fibrous wad placed between the powder and buckshot.) Unless the muzzle of the shotgun is pressed against the victim, he will receive only the buckshot's momentum. The magnitude of the victim's backward momentum will be less than the magnitude of the shooter's because the victim will not be hit by the hot gasses propelling the buckshot out of the gun barrel. Also, thanks to air resistance, the buckshot will be moving slower and have less momentum than when it first exited the gun. If the recoil momentum from discharging a firearm doesn't throw the shooter backward through the nearest window, then certainly the buckshot's momentum won't.

Sometimes Newton's third law is incorrectly evoked as an explanation for why a shooting victim is not blown violently backwards. These explanations claim that the recoil force acting on the shooter and the buckshot's force acting on the victim are an action-reaction pair. To qualify as an action-reaction pair the two forces must

★ Occur simultaneously
★ Be equal in magnitude
★ Be opposite in direction

The recoil force and the buckshot's force fail the first two requirements. The recoil force begins as soon as the buckshot starts moving down the gun barrel. Recoil force is smaller, although it lasts a longer time than the buckshot's force. The buckshot's force does not happen until the buckshot hits its target. The force is very brief but also very large in magnitude compared to the recoil force. There's no way that the force the buckshot creates on the victim and the recoil force acting on the shooter can possibly be an action-reaction pair.

There is one other possible explanation for a victim being blown backward through a window: involuntary muscle contraction. The victim could be so stunned by being shot that he involuntarily tenses his muscles, causing him to jump backward. But it's nearly impossible to jump from a standing position without first bending the knees, which would require one to momentarily relax the leg muscles. Even after bending the knees, it's normally not possible to jump backward by more than 2 or 3 feet (about 1 m). Such a puny jump does not even come close to the distances traveled by Hollywood shooting victims.

RAIL-GUN RECOIL

"They said the physics was impossible" (and it is), yet two bad guys lurk in the shadows watching through impossible x-ray vision scopes mounted on impossible rail-guns as they zero in on their prey, Lee Cullen (Vanessa Williams) and John Kruger (Arnold Schwarzenegger) in the 1996 movie *Eraser* [RP]. The shiny new rail-guns come equipped with LED indicator lights on their sides and green beams of light emitted from their scopes— perfect for those stealthy sniper missions where one must not be seen. When their victims attempt to flee, the assassins send aluminum bullets zinging through the walls at nearly the speed of light, narrowly missing their targets who take refuge behind a refrigerator.

Are the assassins' scopes out of whack? Did they jerk their triggers? Unlike ordinary bullets, bullets traveling near the speed of light would arrive on target almost the instant they were fired. There would be no time for wind, gravity, or motion of the target to create differences between the point of aim and the point of impact. The assassins are within 100 yards (91 m) of their targets. At these distances the misalignment caused by jerking the trigger or the inaccuracies from shooting a shoddy rifle would not usually cause a shooter to completely miss a human-sized target. Even if we imagine that the rail-gun physics make sense, the scene does not.

The rail-gun physics also makes no sense. First, there's the issue of recoil. Yes, even rail-guns would have recoil. It's not necessary to have exploding gunpowder for recoil. The hot gasses from burning gunpowder propelling a normal bullet do add to the recoil, but these gasses exit at about 1.5 times the velocity of the bullet and have far less mass. In a high-powered rifle cartridge such as the 7.62 NATO, the mass of the gunpowder is less than one-third the mass of the bullet. Propelling the bullet at the same speed without using gunpowder would reduce the recoil momentum by about one-third—a significant amount. With a handgun cartridge like the .45-caliber ACP, the gunpowder is only about 4 percent of the mass of the bullet. Here, propelling the bullet without gunpowder would make little difference in recoil. Recoil is caused by sending high-velocity mass out the end of the gun barrel. It does not matter if the mass is a bullet or the gasses from burning gun powder. Likewise, it does not matter whether the force propelling the bullet is from gas pressure or an electromagnetic field.

The big difference in *Eraser's* rail-gun recoil comes from the claim that the bullet travels at nearly the speed of light. Einstein's theory of relativity must be used to calculate the momentum of the bullet at such speeds, making the bullet's momentum far higher than at lower speeds.

If we assume the aluminum bullet exits at 90 percent of the speed of light and weighs 0.26 grams or one-tenth as much as a .22-caliber rim-fire bullet, the backward velocity of the shooter (mass = 100 kg) must be a whopping 1,610 meters per second—about 4.7 times the speed of sound in air—for his momentum to be the same size as the forward momentum of the bullet. Obviously, such recoil is going to impart more than a sore shoulder. It's going to be fatal.

CALCULATING RECOIL WITH BULLET VELOCITIES NEAR THE SPEED OF LIGHT

According to Einstein, when the speed of an object approaches the speed of light, the object's momentum is calculated using the Lorentz factor as follows:

$$p = \gamma m_0 u \quad \textbf{(EQUATION 12.2)}$$

Where:
p = momentum
γ = Lorentz factor
m_0 = mass at rest
u = velocity

$$\gamma = 1/(1 - U^2/C^2)^{1/2}$$

Where:
c = speed of light

For the bullet in *Eraser*:

$$\gamma = 1/[1 - (0.9C)^2/C^2]^{1/2}$$
$$= 2.29$$

Substituting into equation 12.2 yields:

$$p = 2.29 \, (0.00026 \text{ kg}) \, (2.7 \times 10^8 \text{ m/s})$$
$$= 161,000 \text{ N} \bullet \text{s}$$

From conservation of momentum, the backwards momentum of the shooter (ps) must be equal to the forward momentum of the bullet as follows:

$$p_s = 161,000 \text{ N} \cdot \text{s}$$

or

$$m_s v_s = 161,000 \text{ N} \cdot \text{s}$$
$$v_s = \frac{161,000 \text{ N} \cdot \text{s}}{100 \text{ kg}}$$
$$= 1,610 \text{ m/s}$$

The bullet from the magical rail-gun in *Eraser* would also have an unmanageably high kinetic energy—equivalent to several kilotons of TNT. (Note again that Einstein's equations must be used for calculating the kinetic energy.) The bullet is not going to zing through the walls. It's going to demolish the walls, along with the assassins outside and everything else in the immediate vicinity. At distances under 100 yards, the shooter can have his choice: death by the rail-gun's recoil or death by the bullet's impact with the target. At long distance it's only death by recoil.

An aluminum bullet is not magnetic—this point is often cited as a rail-gun physics flaw, but in fact it is not. Such a bullet only has to be conductive. Roughly speaking, a rail-gun works by passing a large current through the bullet from one rail to the other. In the process, a strong magnetic field is generated between the rails at a right angle to the current passing through the bullet. This creates a force acting on the bullet in the direction of the barrel points. Since the bullet is free to move, it accelerates down the barrel and exits as a projectile. The key problem with aluminum is its somewhat-low melting temperature. The electrical current used for acceleration can vaporize the bullet. If the aluminum bullet does make it out the end of the barrel, heat generated at ultra-high velocity by air resistance can also vaporize it or cause it to combust when in contact with air.

Excessive recoil problems can be solved and excessive kinetic energy problems reduced by dramatically lowering bullet mass. But even if the bullet's mass in *Eraser* is reduced to 1/10,000 the size of a .22-caliber rim-fire bullet, the kinetic energy is still going to be the equivalent of over 1,000 pounds of TNT. An aluminum bullet smaller than a speck of dust going at 90 percent of the speed of light will overheat from air resistance and disintegrate the instant it leaves the rail-gun (assuming it doesn't vaporize in the barrel). As it disintegrates, the bullet's kinetic energy will be converted into heat, causing a rather nasty explosion a couple of feet (less than a meter) in front of the shooter's face. There's just not a happy ending available for this device.

In *Eraser*, the rail-gun bullet has huge amounts of momentum that hurls victims horizontally backward great distances. But even here the depiction is bogus. As shown in the movie, the rail-gun bullets are capable of zipping through walls. With such penetration, a human victim would offer almost no resistance to slow the projectile's velocity. Since the bullet has almost exactly the same momentum when it exits the victim, conservation of momentum would still be satisfied even if the victim quietly drops to the floor where he or she is standing. The force of a bullet striking its victim occurs in an extremely short period of time, so it has to be extremely high if it is going to send a person flying backward across an entire room. Such high forces would more likely tear the person apart. Getting hit by a high-energy projectile such as a .50-caliber machine-gun bullet has a gruesome effect on people. It can blow a person in half. It's hard to imagine that a far more energetic projectile from a rail-gun would only throw a person backward.

So how is a victim actually blown backward through windows and across rooms in movies without injury? The actor wears a specially designed harness hidden under his shirt with a rope or wire attached to the back. He is then pulled slightly upward and backward through the window or across the room. The force required to do this is fairly low because it is applied continuously as the person flies backward.

THE COSMIC TOYOTA

It's time to blast off. The hero has been visiting a planet in a distant galaxy (for some heroic purpose) and decides to leave. It's clear that this is an Earth-like planet because the hero can breathe and walk around in just a jumpsuit. The gravity is comparable to Earth and, of course, the planet's higher life forms all speak English. He steps into his spacecraft—about the size of a Toyota—fires the thrusters, and zips off into the cosmos.

Assuming the craft has a mass of 1,000 kilograms, it will take the energy equivalent of about 1,400kg

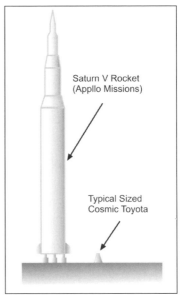

Saturn V Rocket
(Appllo Missions)

Typical Sized
Cosmic Toyota

Figure 19: Saturn V Rocket

(3,100 lb) gallons of gasoline to reach the escape velocity of this Earth-like planet—25,000 miles per hour (40,300 kph), a paltry speed by cosmic standards. None of this, however, is going to strain the cosmic Toyota's fuel supply.

The cosmic Toyota obviously isn't running on gasoline. It has some exotic energy source—say, antimatter. Combining 2.2 pounds (1 kg) of antimatter with 2.2 pounds (1 kg) of ordinary matter would release the energy of about 1.5 billion gallons of gasoline. But there are a few problems. First, storage: antimatter instantly explodes if it contacts ordinary matter. Second, source: there isn't any. Oh well, surely these can be solved in a distant time and galaxy. So, what's the real problem? In a word: thrusters.

Thrusters require an energy source, but they also need a supply of mass to produce the thrust. They work using conservation of momentum. Blast some high-velocity mass out the back of the cosmic Toyota's thrusters, and the craft will gain the same momentum in the forward direction as the expelled mass has in the opposite. The thrust force developed is as follows:

Thrust = (velocity of exhaust) (mass flow rate of exhaust)

Using a very optimistic expelled mass velocity of 50,000 m/s, the Konstantin Tsilokovsky rocket equation predicts that a 1,000 kg Cosmic Toyota will have to expel 800 kg or 80 percent of its mass to reach escape velocity. But the need for mass is actually much worse when blasting off a planet against gravity and air resistance forces. Air resistance does drop to zero outside the atmosphere but is substantial for the first few minutes of liftoff. All these factors together make the mass required for liftoff huge, explaining why NASA needed a behemoth rocket 95 feet 4 inches (29 m) long and 10 feet (3 m) in diameter just to launch the first American, John Glenn, into a relatively low Earth orbit at considerably less than escape velocity. To send three astronauts and their supplies to the moon, NASA required a 363-feet-(111-m-) long 33-feet-(10-m-) diameter Saturn V. Forget energy requirements, a vehicle designed to leave an Earth-sized planet is going to need an enormous mass supply to do so.

As long as it's in an atmosphere, the cosmic Toyota could intake air in the front and exhaust it out the back at a higher velocity, similar to the way a ramjet engine works. Unfortunately, atmospheres tend to be remarkably shallow compared to the distance one must travel to reach escape velocity. So, the supply of mass provided by the atmosphere quickly runs out. Getting a Toyota-sized vehicle off an Earth-like planet using thrusters is impossible without a way to store large quantities of mass.

Once liberated from the Earth-like planet, the cosmic Toyota could cruise around in outer space on a more limited tank of mass, since it would no longer have to overcome a gravity force or air resistance, but it's still going to be a mass hog if its driver is an acceleration freak. If cosmic Toyotas became the rage, the galaxy might become dotted with mass stations. Assuming the cosmic Toyotas came with a few years' supply of antimatter, their thrusters could use just about any form of matter. It might be possible to run them on water or discarded banana peels.

Hollywood could go wild over the opportunities for special-effects scenes. Wreck a cosmic Toyota and its 22-pound (10 kg) fuel tank would explode with the energy of 4.3 nuclear bombs rated at 100 megatons TNT each (the biggest nuclear bombs ever built). An entire disaster movie could be based on a single chase scene and a single wreck.

One could hope that in the faraway time and galaxy heretofore undiscovered, principles of physics or breakthroughs in engineering might solve the problem of the excessive mass needed for escaping from a planet, not to mention the problems of antimatter storage and supply. Sadly, barring such luck, the cosmic Toyota is just too small for shuttling personnel between a planet's surface and an orbiting mother ship, not to mention interstellar travel.

Summary of Movie Physics Rating Rubrics

The following is a summary of the key points discussed in this chapter that affect a movie's physics quality rating. These are ranked according to the seriousness of the problem. Minuses [–] rank from 1 to 3, 3 being the worst. However, when a movie gets something right that sets it apart, it gets the equivalent of a get-out-of-jail-free card. These are ranked with pluses [+] from 1 to 3, 3 being the best.

[–] [–]   Recoil-free rail-guns.

[–] [–]   Shuttle craft the size of Toyotas using thrusters with the ability to escape from an Earth-like planet's gravity.

[–] [–]   Shooting victims blown backward large distances, especially when they crash into glass.

CHAPTER 13

JFK AND MOMENTUM:
Hollywood's Conspiracy to Assassinate History

BACK AND TO THE LEFT

As the open-top limousine cruises silently into view, the president appears to grasp his throat, and is then struck in the head by an assassin's bullet. His head moves back and to the left, its motion captured in shocking detail by the Zapruder film—a bystander's film of the actual events—embedded in Oliver Stone's controversial 1991 movie *JFK* [RP]. During Stone's depiction of the Clay Shaw (Tommy Lee Jones) trial for conspiring to assassinate the president, the scene of Kennedy's head moving back and to the left is repeated again and again as District Attorney Jim Harrison (Kevin Costner) hammers away that it conclusively proves there was a second shooter firing from a position in front of the limousine and not just the lone shooter, Lee Harvey Oswald, firing from behind. His premise is that the victim's head will always be blasted in the opposite direction from the shooter. The physics say otherwise.

ANALYSIS OF KENNEDY'S HEAD MOTION

If the bullet that hit Kennedy in the head had remained embedded in his skull causing no exit wound, Kennedy's head motion would have been easy to analyze with a simple conservation of momentum equation. The analysis would have predicted a forward head motion away from the direction of the shooter. But when the bullet exited the head along with a significant amount of high-velocity tissue, the situation became far more complex. While a simple analysis can no longer be considered conclusive, it can establish if it's possible for the head to move backward instead of forward when struck in the back. We will model the motion as though the head rotated about a pivot at the base of the neck and will use conservation of rotational momentum. A rotational momentum analysis is slightly different than the linear momentum analysis used in the previous chapter, but both obey a conservation of momentum law.

Officially, Kennedy was shot in the back of the head by a 10.37-gram full-metal jacketed bullet fired from a 6.5 × 52-millimeter Italian Carcano WWII surplus military rifle. The bullet's velocity would have been about 552 meters per second at 100 yards distance. We calculate the bullet's rotational momentum as though the bullet is rotating around the base of the neck the instant before it hits as follows:

$$L = I \, \omega \text{ (EQUATION 13.1)}$$

Where:

L = rotational momentum
I = rotational inertia
ω = angular velocity

Assume all objects in the analysis can be modeled as point masses. The rotational inertia for a point mass is

$$I = m \bullet r^2 \text{ (EQUATION 13.2)}$$

Where:

m = mass

r = distance from pivot

Let LB1 = The momentum of the bullet before collision

$L_{B1} = I_{B1} \, \omega_{B1}$

But

$$\omega = v/r \text{ (EQUATION 13.3)}$$

Substitution of 13.2 and 13.3 into equation 13.1 yields:

$L = (m \bullet r^2) \bullet (v / r)$

$$L = m \bullet r \bullet v \text{ (EQUATION 13.4)}$$

$$L_{B1} = 0.01037 \text{ kg } (0.229 \text{ m}) (550 \text{ m/s})$$
$$= 1.31 \text{ kg(m/s)}$$

The military-style bullet that struck Kennedy's head broke into several fragments that cracked the windshield and dented metal trim inside the limousine. Let's assume that the fragments retained 33 percent of the bullet's initial rotational momentum when it exited.
A human head weighs about 11 pounds (5 kg)[15].

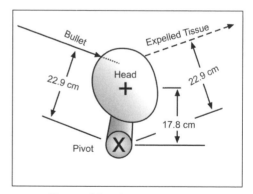

Figure 20: diagram of Kennedy's head

Assume that about 10 percent of the head's mass (or 0.5 kg) of brain, blood, and bone tissue exited the head wound in the forward direction after the bullet exited the head. Also assume that the neck's rotational inertia is minor or negligible compared to the head, and the dimensions of the head are as shown in Figure 20. Zapruder's camera was running at about 18 frames per second, or roughly one frame every 0.057 seconds. The shutter would have been open for about 0.025 seconds during each frame[16] [Zavada, Roland J. "Dissecting the Zapruder Bell & Howell 8mm Movie Camera," http://www.jfk-info.com/zavada1.htm, 10/24/98]. While this sounds like a very short time, it's enough to blur moving objects, making them harder to see. A moving particle, for example, would look like a blurry streak. To calculate an object's average velocity, a researcher would have to estimate how far it displaced and how much time it took to move, then divide displacement by time. But, there would be no good way to know exactly when the motion started. If the motion started halfway through the time the shutter was open, the average velocity would be twice as high as if it had started at the beginning of the shutter opening, assuming the blurry streak was the same length.

Even worse, in every frame, the shutter would close for about 0.032 second so the film could be advanced. No photographic record would be made during this time. These facts alone place limits on the amount of information available for analysis of the bullet's high-speed collision with its target.

There is no evidence of the head shot in frame 312 of the Zapruder film. Frame 313 clearly shows tissue expelled in a forward direction from a head wound. It also shows that a particle was ejected from the wound at an upward angle a distance of over 1.5 meters. It's probably a rotating bone or bullet fragment with at least one reflective side. Its image on film looks like a series of evenly spaced dots (probably corresponding to instances when the reflective side rotated so that it caught the light) connected by a blurry line extending all the way from Kennedy's head. This implies the particle exited the wound after frame 313 had begun. If the particle had exited before the shutter opened, it would have been some distance from Kennedy's head when the photographic record of frame 313 began. There would have been a gap between the President's head and the starting point of the particle's image.

Assuming that 0.5 kg of mass exited the wound and traveled 30 cm (11.8 in) during the 0.025 second the shutter was open in frame 313 gives an average velocity of 12 m/s (26.8 mph or 43.2 kph). The forward rotational momentum of the exiting mass would be as follows:

Let L_m = The rotational momentum of the mass exiting the wound.

From equation 13.4:

$$L_m = m_m \bullet r_m \bullet v_m$$
$$L_m = (0.5 \text{ kg }) (0.229 \text{ m}) (12 \text{ m/s})$$
$$= 1.37 \text{ kg(m/s)}$$

Let L_{B2} = The momentum of the bullet after collision
Let L_H = The momentum of the head after collision

From conservation of momentum:

$$L_{B21} = L_H + L_{B2} + L_M$$
$$L_H = L_{B21} - (L_{B2} + L_M)$$
$$= 1.31 \text{ kg(m/s)} - [0.437 \text{kg(m/s)} + 1.37 \text{ kg(m/s)}]$$
$$= -0.497 \text{ kg(m/s)}$$

Therefore, the head would have had to move backward for conservation of momentum to be true.

But

$$\mathbf{L_H = I_H\omega_H}$$

$$\omega_H = L_H / I_H$$
$$= L_H / (m_H \bullet r_H2)$$
$$= [-0.497 \text{ kg(m/s)}] / [(4.5 \text{ kg})(0.178 \text{ m})^2]$$
$$= -3.49 \text{ radians/s, or } (-200 °/s)$$

In two frames of the Zapruder film, this would have been a motion of about 22 degrees in 0.11 seconds, which is enough to give the perception that the head was being snapped backward (see Figure 21).

Admittedly, the model is too simplistic to be considered conclusive, but it does indicate that a backward motion of the head could be caused by a shot from behind—a possibility that's not even considered in Stone's film.

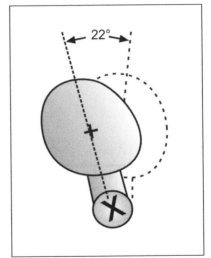

Figure 21: diagram of head motion

Conservation of momentum equations—similar to those demonstrating that a shotgun blast won't blow a shooting victim violently backward—can also show that blasting open a head can make it move in the opposite direction of the bullet's motion (see "Analysis of Kennedy's Head Motion"). In the Kennedy assassination, a significant amount of brain, blood, and bone tissue (there's no way to say it delicately) exited the president's head wound with a forward and to-the-right velocity. This jet of exiting tissue acted like a thruster pushing the president's head back and to the left. In addition, when Kennedy was shot in the back prior to the head shot, he raised his hands toward the exit wound in his throat and elevated his right shoulder causing him to lean slightly to the left. This lean may have assisted the leftward motion of the head.

OK, it takes a lot of simplifications and estimations to make these momentum calculations, and so by themselves the numbers presented above are not conclusive. However, experiments with paint-filled skulls, shown in a November of 1988 NOVA program on PBS, agree that if the bullet passes through the skull and expels fluid at high velocity out the exit hole, the skull will consistently move in the opposite direction of the bullet. Experiments with objects such as melons (Penn and Teller among others have performed this demo[17]), turkey carcasses, and an assortment of other objects have been repeated many times and have shown similar results. Material ejected in the same direction as the bullet acts like a small rocket thruster or jet that pushes the object it was ejected from in the opposite direction. This head motion explanation has become known as the "jet effect."

Normally speaking, exit wounds are significantly larger than entry wounds. Certainly in the Zapruder film, the wound toward the front of Kennedy's head appears much larger than any wound in the back. There are no signs of expelled tissue out the back of the head. This means that if Kennedy were shot in the front of the head, the bullet should have been found in his skull. But if bullet fragments had mysteriously exited from the back of his head with little or no blood spray, they would have fallen in the street since Kennedy was sitting in the back of the limousine. The only bullet fragments found from the head shot were recovered in the front seat area of the car, suggesting that they entered the head from the back and exited from the front, as is consistent with a shot fired from behind.

Still, Kennedy's head motion may have had little to do with conservation of momentum. It may have been nothing more than a random reflex reaction. The analysis of the backward motion of a shooting victim (see Chapter 12) as well as the above analysis shows that victim motion caused by bullets is subtle. Being shot in the head has a tendency to stimulate random nerve impulses. These may have caused Kennedy's neck muscles to involuntarily snap his head back and to the left. The claim that the back and to the left motion of Kennedy's head proved that a second shooter located in front of the limousine fired the head shot is an unsupported conjecture.

How Movies Distort Judgment about Shootings

In films, bullet impacts are generally simulated with "blood-packs," exploded when the victim is supposedly shot. The explosion often sprays a noticeable amount of simulated blood toward the shooter (the opposite direction of the bullet's motion). Exit wound blood splatter is often not simulated. (Why waste a perfectly good blood pack on the victim's back where it's hard to see?) By contrast, in real life there is some blood spray out the entry wound, but the majority of blood and tissue are expelled at higher velocity out the exit wound. On this basis alone, the Zapruder film indicates that Kennedy was shot in the back of the head and the bullet exited the front. But a person whose only knowledge of gunshot effects comes from movies would likely not be convinced by the spray pattern.

Such a person would have seen simulated shooting victims blown violently off their feet and sent flying backward in movies many, many times—pure nonsense according to physics. A movie-indoctrinated person would expect body motion to always be in the direction of the bullet's motion. To this person, any backward motion, such as the backward motion of Kennedy's head, would be convincing evidence that he had been shot from the front. Again, the physics casts doubt.

Could there have been a conspiracy as suggested by the movie *JFK*? Who knows? The motion of the president's head certainly does not support the theory of a second shooter in front of the limousine, which is a key element in the movie's conspiracy theory. Like most things Hollywood, *JFK* seems to have been far more concerned with generating ticket-sale-increasing hype than with presenting insightful analysis.

COUNTING SHOTS (AGAIN)

After over forty years, available forensic evidence still indicates that a lone gunman fired three shots from the fifth floor of the

Book Depository Building, two of which hit President Kennedy. The first hit—now known as the magic bullet—is where much of the conspiracy fun begins. This bullet struck Kennedy in the upper back, exited his throat, and struck Texas Governor John Connelly in the torso, wrist, and thigh. The bullet was later found lying on a stretcher. At first glance, it looks like it's in pristine condition, but closer examination reveals that it is significantly flattened on one side. The bullet was a full-metal-jacketed type, specifically designed for military use requiring maximum penetration with minimal deformation, as required by the Hague Convention of 1899. For humanitarian reasons, this convention banned easily deformed expanding bullets for military use.

The magic bullet lost a large part of its kinetic energy when it passed through Kennedy's neck but had little deformation because it struck no major bones. While passing through Connelly's torso, the bullet glanced off Connelly's ribs, losing more of its velocity in the process. It had tumbled sideways by the time it eventually collided with a major bone in Connelly's wrist; this explains why the bullet was flattened on one side.

Careful analysis of the magic bullet's path by a number of investigators has shown that Kennedy's throat wound does indeed line up with Connelly's numerous wounds. These investigators used photographic evidence for their analysis, including the Zapruder film and 3-D computer graphics simulations[18].

Sophisticated FBI testing with neutron-activation analysis in 1964 and again in the late 1970s showed that the lead alloy in the magic bullet was consistent with lead fragments collected from Connolly's wounds. The testing also showed that bullet fragments found in the limousine matched with fragments taken from

Kennedy's head wound. There's no evidence to suggest that any other bullets struck objects in the limousine. Both the magic bullet and other bullet fragments could be traced to Oswald's rifle.

Yes, some conspiracy buffs still reject all of the above evidence, but to do so usually requires them to claim that the evidence was faked or modified. Many—including Oliver Stone—think the magic bullet was planted. This means conspirators had to fire bullets through Oswald's rifle ahead of time in such a way that the bullets tumbled sideways and were flattened on one side. Confederates with the proper credentials to approach wounded victims would then have needed to wait near possible hospitals for just the right moment and then plant one of the faked bullets. They would have needed to know that none of the real bullets ended up in a recoverable form. Planting an extra bullet would have created all kinds of problems if all the real ones were found.

The additional shooters who supposedly shot Kennedy from the front would have needed disappearing bullets. Any recovered bullet with a caliber other than 6.5 millimeters or micro-scratch marks different from those made by Oswald's rifle would have been a dead giveaway that another shooter was involved. Controlling the final position and condition of a bullet once it's fired would have been problematic at best. If sophisticated conspirators had set Oswald up as the lone assassin, why would they have given him a $12.78 rifle mounted with a supposedly defective $4.58 scope, which according to Stone's movie could not possibly have made the shots? At the time, military surplus semi-automatic M1 carbines mounted with scopes and thirty-round magazines were popular and readily available for under $75—not a huge investment for a group of conspirators. One of these rifles could have spit out at least eight bullets in the time it took Oswald to fire three with his bolt-action rifle. Certainly, an M1 carbine would have been at least as accurate as the 6.5 × 52-millimeter Italian Carcano and far easier to shoot in rapid fire, hence, far less controversial. Had Oswald rapidly fired eight or more shots, witnesses could not possibly have kept an accurate count. It would have been easy to fire a couple of extra shots undetected from a different location, if needed.

Some conspiracy buffs also claim the crime's Rosetta Stone—the Zapruder film—has been deliberately modified as part of a coverup along with x-rays of the president's head, autopsy photographs, and various other forms of evidence. Again, who knows? All this data modification and faking would have required a coordinated effort by all kinds of government groups and individuals, but so what. These are, after all, the same type of groups and individuals who thwarted the Japanese plan to bomb Pearl Harbor, pulled off the Bay of Pigs invasion of Cuba, prevented the 9-11 terrorist attacks, and coordinated the timely hurricane Katrina relief efforts in New Orleans, to name a few known highlights.

How did the Jim Garrison character count shots in the movie? Why, in true Hollywood fashion, of course. According to him there were six, not three, shots fired resulting in seven wounds (he counts entrance and exit points as two wounds). In the movie, Garrison argues against the magic bullet theory using a blatantly incorrect diagram. It shows Connelly seated directly in front of Kennedy at the same height—a position where the so-called magic bullet would have needed to make impossibly sharp turns to have caused the "seven" wounds. In reality, Connelly was seated in a folding jump seat, 3 inches lower and 6 inches to the left of

Kennedy. These seats were used inside the limousine to accommodate the president's entourage but were placed lower so that people seated in them would not block an onlooker's view of the president. Just before he was wounded, Connelly had also turned noticeably to his right.

As for the bullet fragment analysis, the movie has Garrison ranting that

> " . . . the government says it can prove [the single-bullet theory] with some fancy physics in a nuclear laboratory. Of course they can. Theoretical physics can prove an elephant can hang from a cliff with its tail tied to a daisy . . ."

What an insightful comment. Obviously, all forensic experts who use theoretical physics to guide their conclusions are simpletons. So what are they supposed to use? We're never told, but perhaps it's an ultra reliable tool like personal opinion.

The movie's most brilliant analysis occurs when Garrison and an assistant are shown peering out the window of the book depository building overlooking the assassination site. Their dialogue informs us that the Zapruder film established that three shots were fired in 5.6 seconds. (Careful analysis of the Zapruder film indicates that the last two shots were fired within about eight seconds of the first.) The assistant, pretending to be Oswald, then aims, dry-fires, and cycles a rifle identical to Oswald's as Garrison times him. Garrison announces that the time is between six and seven seconds. If you actually time the scene, however, it turns out that the assistant fired the simulated shots in the 5.6 seconds that was supposedly impossible. In a serious moment, comic-magicians Penn and Teller conducted the same dry-firing experiment in under 3.5 seconds[19]. Keep in mind that the clock starts with the first trigger pull, hence, it's only necessary to cycle and aim the rifle twice during the time interval. Naturally, one also wonders about the other three of the six shots Garrison later claimed were fired. If the Zapruder film established that three shots were fired, then what happened to the others?

IT'S ONLY A MOVIE—THE JFK RESPONSE

Inaccuracy in the movie *JFK* goes beyond stupid movie physics, matters of artistic license, or even of interpretation. The movie's problems mired it in controversy even before it was released. It was attacked in the *New York Times, Washington Post,* and *Time* magazine while still in production. Entire Web sites have been devoted to debunking virtually every scene in the film[20]. For example, the movie insists that the president's motorcade route was secretly altered at the last minute to slow the motorcade with a sharp turn and guide the president directly into a killing zone at Dealey Plaza. Yet, the route was not only unaltered but widely publicized ahead of time. It was the only way the motorcade could get from Dallas's Main Street to their desired exit route down the Stemmons Freeway.

The movie centers on Jim Garrison's investigation of an alleged conspiracy to assassinate President Kennedy. Garrison once described the assassination as a homosexual thrill killing and, at various other times, indicated that the CIA, FBI, NASA (yes, the space guys), secret service, Cuban exiles, aerospace industry, Dallas police, Lyndon Johnson, neo-Nazis, and many others too numerous to mention had had a role in the assassination and/or coverup. Of all the suspects he named, he was only able to bring one, Clay Shaw, to trial. The jurors found Garrison's evidence against Shaw so compelling that they took a whopping forty-five minutes to acquit him.

Not to be outdone as a conspiracy theorist, Stone not only tries to convince us that the Kennedy assassination was a conspiracy, but that Lee Harvey Oswald was nothing but a patsy who was unjustly vilified. The movie goes beyond absolving Oswald of guilt as the lone assassin—it actually tries to create reasonable doubt that Oswald murdered police officer J. D. Tippit. When Tippit tried to question Oswald following the assassination, Oswald pulled out a revolver and cold-bloodedly shot Tippit multiple times, finishing him with a shot to the head as he lay helpless on the street. Never mind that it was in broad daylight in front of numerous witnesses; never mind that at least six people picked Oswald out of a lineup and at least two others identified him from pictures[21]; never mind that he had the murder weapon on him when arrested; and never mind that he tried to use it on a second policeman during the arrest—Stone would have us believe that Oswald might have been framed. And what does Stone offer for creating reasonable doubt? Two witnesses who either couldn't or wouldn't identify Oswald, a less-than-sterling job of crime scene investigation by the police, and an unusually large number of police officers showing up for Oswald's arrest. The police presence probably had something to do with the numerous officers already in the neighborhood investigating the Tippit murder, the possibility that many Dallas policemen wanted the distinction of collaring the president's killer, and the radio call that an assassination suspect had entered a nearby theater.

Hollywood responded to the film's controversy by giving it Academy awards for Best Cinematography and Best Film Editing, along with nominations for Best Actor in a Supporting Role (Tommy Lee Jones), Best Director, Best Music: Original Score, Best Sound, Best Writing: Screenplay Based on Material from Another Medium, and, of course, Best Picture. As for the audience, they awarded the film with over $205 million in earnings during its initial movie run.

Naturally, everyone who saw the film dismissed it saying, "It's only a movie." Oddly however, Congress rammed through a bill that formed the U.S. Assassination Records Review Board (ARRB) shortly after the film came out—what a coincidence. The ARRB consumed taxpayer money for about six years and made many government documents on the assassination more accessible to the public, including at least some of the previously sealed materials, but was not mandated to draw conclusions and didn't.

Did Stone's movie unravel a conspiracy, help us understand the assassination, or have any impact whatsoever? Historian Michael Beschloss summarized its impact like this: "the problem that I and most historians would have with Oliver Stone is not his talent—he's a wonderful filmmaker—but that he's used this to put certain myths into the American blood stream that abide to this day."

Summary of Movie Physics Rating Rubrics

The following is a summary of the key points discussed in this chapter that affect a movie's physics quality rating. These are ranked according to the seriousness of the problem. Minuses [–] rank from 1 to 3, 3 being the worst. However, when a movie gets something right that sets it apart, it gets the equivalent of a get-out-of-jail-free card. These are ranked with pluses [+] from 1 to 3, 3 being the best.

[–] [–] Using bad physics as justification for defective historical analysis.

[–] [–] Using Hollywood depictions of shootings as though they were real forensic analysis.

[–] [–] Assuming a shooting victim will always be blown in the same direction that the bullet is moving.

CHAPTER 14

SCENES WITH REAL GRAVITY:
Celebrating Disasters with Happy Hollywood Endings

ESCAPE VELOCITY ON ASTEROIDS

When killer chunks of space junk threaten Mother Earth, what should we do? Dial NASA, slap together a mission, land a spacecraft on the offender, and nuke it. In *Armageddon* it's a Texas-sized rock. In *Deep Impact* a 7-mile-long rock filled snowball. Thanks to gravitational effects, both pose considerable challenges to landing parties.

The daring crew in *Armageddon* had to land on a foreboding spike-covered asteroid. Their choice of landing craft: a space shuttle, of course. At the Kennedy Space Center landing such a craft requires a 15,000-foot-(4,572-m-) long, 300-foot-(91.4-m-) wide specially constructed runway and parachutes for a touchdown at speeds of 213 to 226 miles per hour (343–364 kph). Who knows what the shuttle's speed was in the movie, but a closing speed of over 200 miles per hour was well within reason. The main shuttle boosters were pointed backward, so the shuttle had no means of slowing down upon landing other than applying brakes or running into rock formations. Certainly, parachutes would have been worthless since there was no atmosphere. Airliners touching down on smooth water at similar speeds routinely disintegrate, but not— by golly—our hope-of-all-humanity space shuttle when landing on misshapen rock formations.

Once on the asteroid, crew members walked around normally in spite of the reduced gravity. In fact, gravity would have been about 10 percent of the level on Earth, assuming the asteroid's density was the same as Earth's. By comparison, the moon's gravity is only 17 percent of Earth's—an amount less than the smaller-sized asteroid due to the moon's lower density.

G-FIELD BASICS

According to Newtonian mechanics, any mass, such as a planet, creates a gravitational force field around it. The strength of the force field is usually represented by the symbol g, generally referred to as the acceleration due to gravity. Although g has units of acceleration, an object's acceleration in a gravitational field will only be equal to g when it is free falling. The gravitational force acting on an object will be mg (where m is the object's mass) regardless of whether the object is freefalling with an acceleration of g or sitting on the ground with an acceleration of zero. It's more accurate to refer to g as the gravitational field strength.

GRAVITATIONAL OR WEIGHT FORCE ON A PLANET, MOON, OR ASTEROID

To find the gravitational force acting on an object, in other words its weight, multiply its mass times g as follows:

$$F = mg \text{ (EQUATION 15-1)}$$

Where:

F = force due to gravity

m = mass of the object in a gravitational field

g = the strength of the gravitational field (g = 9.8 m/s^2 for Earth)

GRAVITATIONAL FIELD STRENGTH—AROUND A PLANET, MOON, OR ASTEROID

For any spherical-shaped celestial object, the gravitational field strength g at a specific location on or above the surface can be calculated as follows:

$$g = G \bullet M_P / r^2 \text{ (EQUATION 15-2)}$$

Where:

G = the universal gravitational constant

M_P = the mass of the planet

r = the distance from the center of the planet

Note that if the distance from the center of the planet is doubled, the gravitational force decreases by a factor of 4. Gravitational field strength drops quickly as one moves away from the planet.

GRAVITY FIELD STRENGTH—INSIDE A PLANET

If a shaft is drilled into the interior of the planet, g can be found as follows (assuming that the density of the planet remains constant):

$$g = G \bullet M_P \bullet R / R_P^3 \text{ (EQUATION 15-3)}$$

Where:
R_P = the radius of the planet

Note that the gravity level actually increases as one moves away from the center of the planet, that is, until reaching the surface. Then, at the surface, equation 15-2 takes over and the gravity level starts dropping.

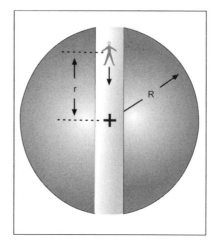

Figure 22: inside of a planet

The movie at least has one of the two shuttles crash, but, of course, all hands are not lost. The surviving crew members set out in a rover hoping to join up with members of the intact shuttle. Faced with crossing a deep chasm, they simply rev up the rover and roar over the side. With the low gravity and no air resistance, the rover supposedly goes into a close-to-surface orbit, easily crossing the chasm without falling into it. If an object close to the surface has no horizontal velocity and is dropped, it will fall straight down. Give it a horizontal velocity, and it will fall in a parabolic arc. Give it enough horizontal velocity, and it will "fall" in a stable circular orbit. Yes, such an orbit is possible on the surface of a planet or asteroid and requires a much lower horizontal velocity on the asteroid than on Earth. On our Texas-sized asteroid, however, the rover would have to travel at a speed of 1,800 miles per hour (2,900 kph) to reach the required critical velocity—a little fast for the typical rover.

Astronauts on the surface of the Texas-sized asteroid could jump up and down as much as they wanted with no danger of reaching escape velocity and accidentally floating off into space. Escape velocity is proportional to the square root of the gravitational field strength and, surprisingly, the gravitational field of the asteroid would be a little stronger than on the Moon, assuming the asteroid's density matched that of Earth rather than the lower value of the moon. On Earth escape velocity is slightly over 25,000 miles per hour (40,000 kph). On the asteroid it's greatly reduced but still over 2,500 miles per hour (4,000 kph). Taking off from the surface in a horizontal direction would make reaching escape velocity a real challenge.

CRITICAL VELOCITIES RELATED TO GRAVITY

There are two critical velocities related to gravity that determine how and if an object orbits a celestial body. (A celestial body in this case refers to a spherical body like a planet.)

LOWEST CIRCULAR ORBIT VELOCITY

An object could conceivably orbit a planet a fraction of a millimeter off the surface, if the planet were perfectly spherical, uniform in density, and had no air resistance to slow the object's speed. The horizontal velocity required for the lowest possible circular orbit defines the first critical velocity. This velocity is perpendicular to a radial line drawn from the center of the celestial body.

If a spacecraft attempts to take off in a horizontal direction on a celestial body that has no atmosphere, the craft's velocity will have to exceed the first critical velocity. Without a lengthy well-maintained runway, a horizontal takeoff is next to impossible, which is why missions to asteroids, moons, or planets with thin atmospheres will require vertical takeoffs and landings. On Earth, airplanes can take off at relatively low horizontal veloci-ties thanks to the atmosphere, which provides lift.

In a circular orbit, the force of gravity acting on an object acts as the centripetal force. The velocity required for a circular orbit is calculated as follows:

$$V_{c1} = (G \bullet M_P / r)^{1/2} \text{ (EQUATION 15-4)}$$

Where:
G = the universal gravitational constant
M_P = the mass of the planet
r = the distance from the center of the planet

The first critical velocity V_{c1} would be found by substituting the radius of the planet rp as follows.

$$V_{c1} = (G \bullet M_P / r_P)^{1/2} \text{ (EQUATION 15-6)}$$

ESCAPE VELOCITY

Escape velocity is the second critical velocity value. This is the minimum velocity required to escape from the gravitational pull of the celestial body. If an object is moving with a tangential velocity between the first and second critical speeds, it will move in an elliptical orbit.

The escape velocity is derived from energy equations. The work required to move an object from a planet's surface to infinity is set equal to the object's kinetic energy. This results in the following expression:

$$V_{c2} = (2G \bullet M_P/\ r)^{1/2} \text{ (EQUATION 15-5)}$$

Note that although the escape velocity is derived in a different manner, it's simply the circular orbit velocity multiplied by the square root of 2. When a space vehicle slingshots around a planet or moon, it has to be going at or above the escape velocity to avoid being captured in an orbit or—worse—crashing into the surface (if the vehicle is below the circular orbit speed). Slingshotting allows a spacecraft to make a turn without using fuel. Normally, turns consume large amounts of fuel.

SUMMARY OF THE EFFECTS OF A TANGENTIAL VELOCITY AROUND A PLANET

★ First critical velocity—the lowest velocity possible for taking off from a celestial body with no atmosphere.
★ Second critical velocity—the lowest possible velocity for escaping a celestial body's gravity.

If the booster motors were vectored with a downward angle, they would push the shuttle's tail upward. Unfortunately, this would also rotate the shuttle's nose downward. Yes, the shuttle would have small thrusters in its front, which in theory could counteract some of the nose's downward rotation. But they would be low-powered devices used for positioning the shuttle when in orbit. They would not be designed for assisting a horizontal liftoff. Keep in mind that on the asteroid the gravitational attraction force on an empty space shuttle would be over 16,000 pounds (7,300 kg). With no air, there would be no lift generated by the shuttle's wings. Getting the shuttle to rise off the surface using booster motors pointed in a horizontal direction would be a major feat.

There are two critical velocities associated with gravity: the velocity required for a circular orbit at the surface and the escape velocity. If the shuttle is moving horizontally with no lift force, it would have to exceed the velocity required for circular orbit at the surface before it could hope to lift off. As mentioned earlier in the rover discussion, this velocity is 1,800 miles per hour (2,900 kph)—a new land-speed record for humanity if not for Earth. Once off the surface, the shuttle is going to have to ramp up its speed to over 2,500 miles per hour (4,000 kph) in order to escape the asteroid's gravity.

Even if the shuttle did miraculously manage to leave the spike-covered asteroid, it would likely not survive reentry into Earth's atmosphere. The space shuttle Columbia burned up during reentry due to damage sustained when its wing hit a chunk of foam insulation that broke off its fuel tank during launch. How are a shuttle's heat tiles going to survive scraping against and smashing into an asteroid's rock formations?

Deep Impact was far more realistic than *Armageddon*. Its 7-mile-long comet would have had only about 0.09 percent of Earth's gravity level, and the movie portrayed it that way. Astronauts on the surface were attached to tethers to keep them from floating off. Even here, a person could not easily reach the escape velocity of 22 miles per hour (36 kph) simply by jumping. But escape velocity was still way too low for comfort.

The astronauts touched down on the comet in a specially designed lander rather than bulldozing a landing strip with a space shuttle. Naturally, nuclear bombs had to be drilled into the comet's surface not at 75 feet, but at exactly 100 feet. This drilling had to be done at some distance from the landing craft so that the craft could almost run out of fuel as it raced to pick up the tardy drilling crew—just in the nick of time. Even then, the comet had rotated into the sunlight, causing an astronaut to be blown into space by a violent out-gassing of comet material instantaneously vaporized by the sunlight's heat.

It's not that the astronauts were dawdling. The digging device got stuck, and a brave astronaut had to selflessly climb down the 75-foot-deep hole then jump up and down on the digger to get it working. This took longer than the allotted time, but a bomb was successfully planted.

Back on the ship, the bomb was detonated. After all the bravery, the ingenuity, and the close calls, the mission failed. The bomb was supposed to merely alter the course of the comet. Instead, the explosion split the comet into two pieces—a big and a little one. Both of these parts continued on a path toward Earth, with the little guy racing ahead of the big one.

The split comet was a nice touch that illustrated one of the problems involved with trying to save Earth from an impact disaster. For the small piece to remain separated from the large one and speed ahead of it, the small piece would have needed to reach escape velocity, which would have been less than 22 miles per hour since the large piece had less mass after the small piece broke off. To reach such a velocity, the small piece would have needed to pick up the kinetic energy equivalent of about 10 megatons of TNT (assuming the small piece was 20 percent of the comet). The total explosive energy of the eight 2-megaton nuclear bombs carried by the spacecraft would have been marginal for providing it. Failing to reach escape velocity, the little piece would have been pulled back to the larger piece. Okay, it probably would not completely fuse back together. And maybe it would still be possible to shatter the big piece at the last minute by crashing the space ship into it as depicted in the movie. Still, if a nuclear bomb planted on the surface couldn't effectively break up the comet, it's doubtful that crashing a spaceship loaded with still other nuclear bombs of the same size would do much better.

KILLER TIDES

Let's digress back to *Armageddon* and assume success in spite of all the space shuttle problems. It's a momentous occasion. The Texas-sized asteroid on a collision path with Earth has just been split in half. The parts pass on either side of the Earth within a mere 400 miles of the surface. But why are people joyful? The oceans would have sloshed out of their basins and sent walls of salt water smashing across the world's coastal areas. Places like Florida would be submerged. The water walls would be so heavy they would destabilize fault lines, setting off earthquakes and volcanic eruptions. When the oceans finally quit sloshing back and forth and the water receded, the sea in coastal areas would be filled with all kinds of sediment and contaminants. On land in coastal areas, dead marine life and people would be scattered everywhere, water systems contaminated, crops destroyed, and major cities demolished—not a cause for celebration.

Tides are normally created by shifting gravity forces that the Moon and, to a lesser extent, the Sun exert on the oceans. The mass of one-half of the asteroid would be about 3×10^{21} kilograms as compared to 7×10^{22} kilograms for the Moon. In other words, the moon is about 23 times more massive than the asteroid half, and gravity forces are directly proportional to mass. So what's the big deal? Unfortunately, the gravity force is also inversely proportional to the square of the distance between the centers of mass of the objects causing them. The distance between the Earth's center and the moon's center is about fifty times longer than the distance between the Earth's center and the asteroid half's center. Taking into account all of the differences, the gravity force acting on Earth caused by half an asteroid is almost one hundred times higher than the gravity force on the moon. But there are two of these forces, one acting on each side of Earth. From the standpoint of tides these tend to reinforce each other. Keep in mind that ordinary tides caused by the Moon are around 10 feet (3 meters) and take several hours to rise. The asteroid-produced tides would rise much higher in much less time.

Water, however, is only one source of devastation. There's also wind. Normally the atmospheric tides created by the Moon are so small that they have almost no effect on weather. But increase these forces by a factor of 100 on opposite sides on the globe, and the result would be high winds over the entire surface of Earth. Underneath the passing asteroid halves, the winds could easily act like continent-sized hurricanes, adding major-sized storm surges to the already incredibly high tides.

The asteroid pieces would pass around Earth and collide back together on the other side in about half a day. They would then fly off into the cosmos. The gravitational pull of the asteroid mass would diminish quickly as it moved away from Earth. Within just a few hours the pull would be negligible. Unfortunately, the disastrous problems on Earth would persist for some time. The back and forth sloshing action of the ocean would last for hours if not days. High winds and erupting volcanoes with sloshing oceans could disrupt weather patterns for decades if not centuries, not to mention that such a large mass passing close to Earth and the moon could disrupt their orbits with unknown consequences. Earth would be a mess.

THE DIRECTION AND STRENGTH OF "INNER" GRAVITY

Imagine digging a shaft from the surface of Earth all the way through the center and out the other side (sounds like a movie plot already). Of course, there would be some minor problems such as extreme heat and pressure, but suppose these could be overcome. If a group of people descended into the hole in some type of elevator and stopped every so often to measure the force of gravity, they would find that it slowly decreased to zero at the center of the world and then increased back to its normal level as the elevator passed the center and ascended to the

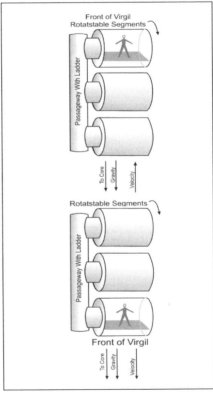

Figure 23: Virgil

surface on the opposite side. If the density of Earth were constant, gravity force would be directly proportional to the distance from the center of Earth. When the inner core was reached, the gravity force would be one-sixth of its value at the surface. *The Core* does not depict this reduction in gravity force as the crew of the *Virgil* bores down toward Earth's core. The inner core, however, is 4.5 times denser than the surface, and so the gravitational attraction force at the inner core would only be reduced by around 25 percent—not enough difference to mention.

Where the movie fails is in its depiction of gravity's direction. Until the center of Earth was reached, the force of gravity would always be downward. Yet the *Virgil* is configured like a vertical subway train. Its crew walks around in their ship as though the gravity force were rotated by ninety degrees. They never have to climb ladders to go from the front of the ship (its lowest point) to the back (its highest point). As the ship passed its lowest depth inside Earth and started heading back out, the gravity force direction should have flip-flopped, but this was also not depicted.

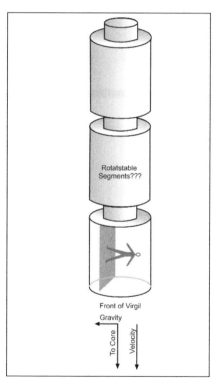

Figure 24: Virgil as depicted in the movies

When the *Virgil's* crew was in training for the mission, the movie indicated that the crew compartments could be rotated so that gravity remained pointed in the right direction. But this still did not solve the problem of moving from compartment to compartment. The design would have to be more complex than the simple subway-type design depicted in the movie. The movement between compartments would have to be done by climbing up and down ladders inside tubes connecting the pieces. The connecting tube would have to be located to one side of the compartments like a backbone so that each compartment could be rotated 180 degrees. Clearly, the moviemakers did not waste time working out such details. In a sense they had a point: why worry about minor details like the direction of gravity when the whole premise of the movie was ridiculous.

Summary of Movie Physics Rating Rubrics

The following is a summary of the key points discussed in this chapter that affect a movie's physics quality rating. These are ranked according to the seriousness of the problem. Minuses [–] rank from 1 to 3, 3 being the worst. However, when a movie gets something right that sets it apart, it gets the equivalent of a get-out-of-jail-free card. These are ranked with pluses [+] from 1 to 3, 3 being the best.

[–] [–] Assuming that a near miss by a gigantic asteroid would cause no harm to Earth.

[–] [–] Paying no attention to the direction of gravity.

[–] [–] Underestimating or overestimating the gravity levels, surface orbit velocity, and escape velocity of an asteroid.

[–] [–] Space shuttles landing in impossible situations without damage.

[–] [–] Space shuttles taking off in impossible situations.

CHAPTER 15

SCENES WITH ARTIFICIAL GRAVITY:
The Good, Bad, and Ugly Space Stations

THE VOMIT COMET

The person who said, "crime doesn't pay," never watched a space movie. While some movies are upstanding, most are law-breakers—laws-of-physics breakers, that is—especially when it comes to gravity. Trying to simulate the lack of or apparent lack of gravity takes creativity and hard work[22]—an anathema to the prospects of easy money.

Apollo 13 [GP] (1995) pulled the simulation off beautifully using NASA's "Vomit Comet" (the KC-135A aircraft similar to the Boeing 707 commercial airliner). During a parabolic dive cycle, this aircraft provides about twenty-five seconds of apparent weightlessness but unfortunately cycles between the sensation of zero and elevated gravity—a cycle that drives the inner ear bonkers, producing mild to extreme nausea, hence, the nickname Vomit Comet. When the *Apollo 13* movie crew needed to film a space flight scene, they didn't do lunch, they risked redoing lunch. The result, however, was worth the indigestion: *Apollo 13* was the third highest grossing film in 1995, with two academy awards and seven additional academy award nominations.

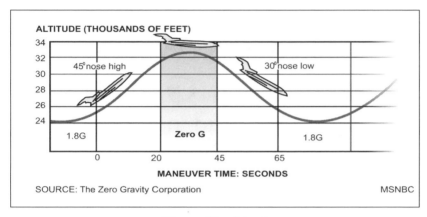

Figure 25: altitude

At the start of a parabolic cycle, the Vomit Comet climbs upward at a forty-five-degree angle. As it rounds the top section of a parabolic arc, the aircraft and all its contents are in freefall, which feels like zero gravity. The sensation is sometimes incorrectly referred to as zero-gravity, but, in reality, it's not: the gravity force is just as high as it would normally be. The condition is also sometimes referred to as zero gs, but gs are a unit of acceleration and the acceleration of the aircraft is 1.0 g downward, even when simulating weightlessness. The sensation occurs only because the aircraft is in freefall.

As the aircraft heads toward the ground, the pilot has to pull it upward and the sensation of gravity returns with a vengeance. At the bottom of the dive a person will feel as though they are 80 percent heavier.

CREATING ARTIFICIAL GRAVITY

For filmmakers with finicky stomachs, artificial gravity is the cure. It requires no unsettling parabolic cycles, but to do it right requires some understanding of physics. Keep in mind that artificial gravity is not really gravity at all. To understand it, we first must understand which force causes the sensation of weight for a person standing on the ground. It's not the gravity force. It's something called a normal force.

Obviously, there's a gravitational force pulling the person downward, but since the standing person is not moving or sinking into the ground, there must be another force of the same size pushing the person upward. The two forces cancel each other out. The upward force is called a normal force and is simply the upward force the ground creates on the person. Okay, the idea of the ground pushing upward may

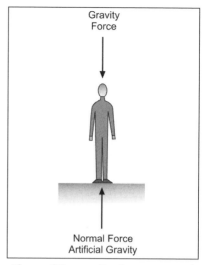

Figure 26: gravity force normal

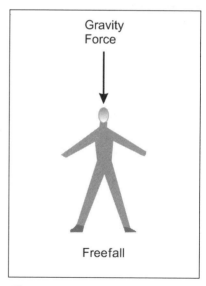

Figure 27: gravity force freefall

seem outlandish, but it does. The person's feet push downward on the ground; the ground pushes upward on the person. It's an action-reaction pair. The normal force always acts perpendicular (or normal) to the surface that creates it, hence the name normal force.

Since there are actually two forces acting on a person standing on the ground—a gravity force and a normal force—one, the other, or both must cause the sensation of weight. Let's run a little mind experiment to find out which it is. First, let's spring a trap door on a person and cause him to fall into a bottomless pit. In other words, remove the normal force but leave the gravity force. Other than the emotions from falling down a bottomless shaft, how does the person feel? Weightless. In fact, anyone in free fall will feel weightless. Clearly, the normal force is required for having a sensation of weight.

THE BASIS OF ARTIFICIAL GRAVITY

Artificial gravity has nothing to do with gravity. In reality, it is the normal force acting as a centripetal force in a rotating cylinder, disk, or doughnut-shaped spacecraft that produces the effect. The normal force pushes upward on the inhabitants standing on the rotating floor. It is derived as follows:

$$F_c = ma_c$$

Where:

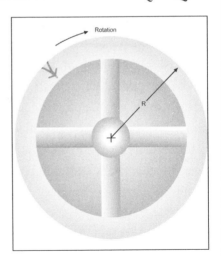

Figure 25: rotation

F_c = centripetal force
m = the mass of the object or person subjected to the artificial gravity
a_c = centripetal acceleration

Centripetal acceleration is calculated as follows:

$$a_c = v^2/r$$

Where:
v = the object's tangential velocity
r = the distance from the center of rotation to the object's center of mass

The centripetal force or normal force must be equal to a person's weight on Earth to make the person feel like he or she is in an Earth-like gravity field. Hence, the centripetal acceleration must be equal to 1.0 g (9.8 m/s²). In equation form:

$$g = v^2/r$$

Solving for velocity gives:

$$v = (gr)^{1/2} \textbf{ (EQUATION 15.1)}$$

However, it's inconvenient to talk about rotation in terms of tangential velocity. Revolutions per minute (RPM) is a more common way to quantify rotation.

Assume that v is in units of meters per second and r is in units of meters. The object travels the circumference of the circle ($2\pi r$) in each revolution. Then the number of revolutions per second (RPS) is as follows:

$$RPS = v / (2\pi r)$$

Substituting for v using equation 15.1 yields:

$$RPS = (gr)^{1/2} / (2\pi r)$$

$$RPM = 60 \; (gr)^{1/2} / (2\pi r)$$

For Earth g = 9.8 m/s2

$$RPM = 30 \; (9.8)^{1/2} / \pi r^{1/2}$$
$$RPM = 30 / r^{1/2}$$

For the 400-meter diameter space station in *2001: A Space Odyssey* the rotational speed would be calculated as follows:

$$RPM = 30 / (200 \;)^{1/2}$$
$$= 2.12$$

Next, let's pull the imaginary giant gravity switch in the sky to the "off" position. Yeah, yeah, it sounds like something in a movie, but doing physics often requires imagination. Our person again feels like he's weightless. So, it seems that a gravity force is also needed. But wait. When the gravity force was turned off, the normal force also ended. Unfortunately, we still don't know if we need gravity to have a sensation of weight. To answer the question we have to create an experiment with a normal force but no gravity force. If the person still feels the sensation of weight, then it's obvious that a normal force is required but a gravity force is not.

To run this experiment, we can have subjects stand against the wall inside an amusement park ride—the kind that is a large hollow cylinder that spins around a vertical axis at its center. When the ride is spinning at full speed and the riders are pressed against the cylindrical wall, the bottom drops out. This results in some screaming, but no one falls out. They feel like they have a large gravity force pressing them against the wall so

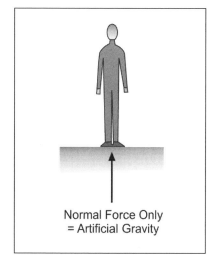

Figure 29: normal force only

firmly that they can't slide out. But there is no gravity force in the horizontal dimension. What they are feeling is the normal force between the cylindrical wall and the back sides of their bodies. This force acts as the centripetal force that makes the subjects speed around in a circular path. A normal force acting without a gravity force can indeed produce the sensation of weight. The conclusion: normal force, not gravity, is responsible for the sensation of weight.

Okay, at this point someone invariably says "but what if a person is hanging from a cable? There is no normal force." The answer is that a normal force is simply the force a surface exerts on an object—in this case a person. To hold the person up, the surface of the cable or the surfaces of a harness attached to it must be in contact with the person. Add up all the normal forces from the cable or harness, and it will equal the force of gravity but act in exactly the opposite direction. In other words, the normal force does still exist and is indeed the primary force causing the sensation of weight.

Someone will invariably also argue that the amusement ride described above does not turn off the force of gravity. However, the sensation of weight described was in the horizontal dimension corresponding to the normal force exerted by the cylindrical wall. Gravity is in the vertical dimension. Seal the ends of the cylinder, provide it with a life support system, transport it to outer space, spin it up to speed with the people against the wall, and they will still feel a weight-like sensation.

If the cylinder has a large diameter—say 440 yards (400 m) so that it looks disk shaped—and is spinning at exactly the right speed, the people will be able to stand on the cylindrical wall and walk around just like they were standing on Earth. If one of them drops a coin, it will appear to fall to the floor (the cylindrical wall) just like a coin would on Earth. The effect is called artificial gravity. In reality it is an artificially produced normal force. There is no gravity force involved. The normal force is providing a centripetal force in the direction of the center of the cylinder. This centripetal force keeps people rotating around inside the disk. When a coin is dropped, the centripetal force is momentarily removed and the coin falls to the floor, which then reestablishes the centripetal force.

The rotating cylinder needs to have a large diameter for two reasons: it makes the floor's curvature less noticeable, and it makes the artificial gravity effect on the person more uniform. With artificial gravity the centripetal force provides the artificial normal force needed for the sensation of gravity. Unfortunately, centripetal force depends on one's distance from the center of rotation, or in this case, the center of the disk-shaped space station. It will be zero at the center and 1.0 g at the floor people stand on. (Keep in mind that the floor is the cylindrical wall of the disk.) If the space station had a diameter of 12 feet (3.7 m) and was rotating fast enough so that objects on the floor would experience 1.0 g of acceleration, six-foot-tall people standing on the rotating floor would experience 1.0 g in their feet and 0 g in their heads. The result: who knows? But nausea, light-headedness, and disorientation would be good bets, especially if a person bent over, sat down, or stood up.

THE GOOD PORTRAYAL OF ARTIFICIAL GRAVITY

A realistic space station with artificial gravity would look like a gigantic rotating bicycle wheel with large spokes—just like the space station in *2001: A Space Odyssey* [GP] (1968). Yes, *2001* has some out-there stuff, such as the monoliths, which have about as much scientific basis as magic wands, not to mention the movie's colorful but incomprehensible ending (reading the book was required to make any sense of it). But when a scene calls for realistic Newtonian physics, the movie delivers. Its realism created a mood that has never been duplicated.

2001 won an academy award for visual effects, is listed number twenty-two in the American Film Institute's 100 Greatest Movies list (http://www.afi.com/tvevents/100years/movies.aspx), made $56.7 million (270 million in 2004 dollars), and developed a cult following; and yet it ranks as one of Hollywood's least copied movies. If influence over other films were measured in height above sea level, *2001* would be in the Marianas Trench. Why? Hollywood has yet to understand its success let alone develop a formula for cloning it. Still, a few aspects of the movie have stuck, such as the vague notion that something has to rotate to produce artificial gravity.

THE BAD PORTRAYAL OF ARTIFICIAL GRAVITY

To give itself credibility, *Mission to Mars* [RP] (2000) had its spaceship rotate to produce artificial gravity and its inhabitants correctly walk around on the inside of its outer wall. The ship, however, was only about 24 feet in diameter (7.3 m). When standing on the floor in this ship, the feet of people six feet tall would experience 1 g while their heads would experience 1/2 g. If they sat down, the acceleration of their heads would increase by 50 percent. When they climbed the ladder leading to the center of the ship, their centripetal acceleration would have decreased to near zero, yet none of these changes produced even the slightest nausea.

THE UGLY PORTRAYAL OF ARTIFICIAL GRAVITY

Armageddon [RP] (1998) took the foolishness even further. They had an astronaut press a button in a Mir-type Russian space station (mass = 124,340 kg) and spin it up to full artificial gravity level in a matter of seconds. The stresses from the high acceleration required to spin up so quickly would have torn the space station apart. Inside the space station the distance from the axis of rotation varied from 39.4 feet (12 m) to as little as 6.6 feet (1 m), yet the gravity level was always the same in any part of the station. Both the station inhabitants and its visitors experienced no discomfort when their heads were located at 0 g and their feet at 1.0 g. The artificial gravity was so remarkable that it always acted in the correct direction even when the visitors were climbing through tubes pointed directly away from the axis of rotation.

Red Planet [RP] (2000) takes the award for most muddled understanding of artificial gravity. The spaceship in it looks like an oversized shoebox with large counterrotating rings on each end that seem like they belong on a carnival ride with neon lights. While it's not completely clear, the main crew quarters seem to be located in the nonrotating section of the shoebox. So it's a mystery how the crew can walk around inside it just like they were on Earth.

The box-like shape of the central ship also makes no sense—assuming it's pressurized for human habitation. The internal pressure inside it would tend to make the sides bulge outward. A sphere or a cylinder with spherical ends would be a far better shape for containing internal pressure.

Having the rings counterrotate with a stationary shoebox in between means the rings must have large-sized bearings and seals to prevent air loss. Thanks to seal friction, the rings would have to constantly be driven by electric motors—a power drain on long space voyages. Compare this to the elegance of the *2001* space station in which the entire space station rotates. Once it's turning, it will rotate indefinitely with no further power input. There is no air resistance or friction in space to slow it down, nor are there any bearings or seals required.

When a solar flare disables the *Red Planet* spaceship, the counterrotating rings stop moving, causing the artificial gravity to fail. The rings on the ship would have a huge amount of rotational inertia, and yet they are brought to a halt in seconds. The accelerations required to do this would likely tear the rings apart. A large spacecraft would likely not be designed for unexpectedly high stresses, since every pound of material used to build it would cost thousands of dollars just to lift it off Earth and send it into space for assembly.

As the rings stop, loose objects aboard the ship begin to float about. But why? They would still be moving in the direction of rotation. Quickly stopping the rings should send objects tumbling into the nearest wall in the direction of their motion—that is, if the crew quarters were correctly set up in the rotating rings.

When the rings are again restarted, artificial gravity is instantly restored and all the randomly floating objects immediately drop to the floor. If they really were gently floating inside a stationary ring, the objects would also be stationary. When the ring started rotating, the wall opposite of the one they had previously tumbled into would appear to move forward and bump into all the floating objects. As the rings accelerated the objects would slide toward the floor and pile up on top of each other.

Artificial gravity on the starship *Enterprise* in the *Star Trek* movie series is beyond the categories of good, bad, and ugly because no attempt is made to use centripetal force as the source. *Star Trek* proposes that humanity is so advanced, a spaceship can be built that manipulates gravity in a similar way as it manipulates other phenomena such as electricity—once considered esoteric. The series offers no mumbo-jumbo explanations. If the *Enterprise* couldn't generate its own gravity, the entire movie series would have to be rethought. The stretch is forgivable but in a begrudging manner. Had the writers used reliable Newtonian physics to generate artificial gravity, the *Enterprise* would have looked different and the stories been altered, but it still would have had the *Star Trek* flavor.

At first glance, a rotating ring like the huge *2001* space station flying through space sounds weird, but why not? As long as the vehicle is neither required to land nor take off, there's no reason for making it aerodynamic. It will encounter no air resistance. The ship will have no fuel-guzzling liftoffs, and once it's up to cruising speed it will need no energy to keep it there. Hence, small size is not critical. Such a ship would still have a great deal of screen presence.

Robert Zubrin's highly creative proposal for a Mars ship would break all cinematic paradigms in spaceship design. He suggests that a section of used-up booster could be tethered several hundred meters from the living quarters of the Mars ship and the two rotated around a point in the middle. In a military version both sections could hold living quarters. Imagine the spectacle of an invading army in hundreds of tethered ships spinning as they silently traveled through space. Realistic artificial gravity does impose design constraints on cinematic spaceships, but even in the world of reality there's still plenty of room for creativity and gee-whiz effects.

Summary of Movie Physics Rating Rubrics

The following is a summary of the key points discussed in this chapter that affect a movie's physics quality rating. These are ranked according to the seriousness of the problem. Minuses [–] rank from 1 to 3, 3 being the worst. However, when a movie gets something right that sets it apart, it gets the equivalent of a get-out-of-jail-free card. These are ranked with pluses [+] from 1 to 3, 3 being the best.

[–] [–] Depictions of spacecraft in which rotating parts produce artificial gravity (AG) but the part of the craft with the gravity does not rotate.

[–] [–] Depictions of AG in which the direction of the AG does not match with the direction of centripetal force. (Note: centripetal force always points at the center of rotation.)

[–] [–] Depictions of AG in which peoples' heads are subjected to totally different AG conditions than their feet.

[–] [–] Starting or stopping rotation in massive spacecraft parts in a matter of seconds.

[–] Objects falling straight to the floor when interrupted AG produced by rotation is restored.

[+] Astronauts becoming dizzy and possibly vomiting when they climb ladders taking them from high rotationally produced AG to low AG.

[+] [+] Using a Zubrin design for producing AG.

CHAPTER 16

THE MOVIE MERRY-GO-ROUND:
How Filmmakers Create Ridiculous Spin

MARTIAL ARTS

It's possibly the most famous movie kick ever. Trinity leaps in the air, arms extended like the wings of a white crane about to fly. For a moment time stops. When it resumes, she falls toward Earth slamming her foot forward and slightly downward into a fat cop's chest. He flies horizontally backward across the room and crashes into a wall (*The Matrix* [RP]). The action does take place inside a computer simulation, and Trinity has supposedly mastered altering the simulation's physics, but what about the fat cop? Surely, he hasn't. He's part of the simulation and should have rotated backward and crashed into the ground a short horizontal distance from where he got kicked.

Increasing the power of the kick would cause the hapless cop to slam into the ground harder, but would have little effect on the distance he flies backward. In fact, at some point the extra power will mostly go into crushing the rib cage and collapsing the lungs rather than increasing the victim's kinetic energy.

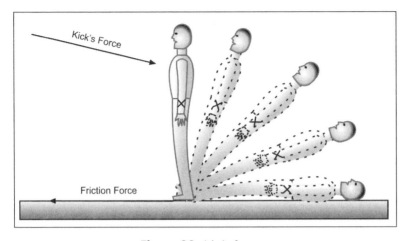

Kick's Force

Friction Force

Figure 30: kick force

The only way a victim can be sent straight backward is to apply a horizontal force through or, in other words, lined up with the victim's center of mass. Even so, the force must have a slightly upward direction to knock the victim's feet off the ground, thereby eliminating the friction force between the victim's feet and the floor. Since Trinity's kick does not act through the center of mass, the friction force would otherwise cause backward rotation.

The force from Trinity's kick obviously did not match the conditions required for translation (moving in a linear fashion), so why did the cop fly such a long horizontal distance? Run the scene in slow motion, and it's possible to see the glint of a horizontal wire pulling the hapless cop backward. Indeed, the laws of movie physics always require the hapless victim to fly backward—if he or she is going to be thrown horizontally at all—because slamming a stunt person front wise, even into a fake wall, tends to be hard on the nose, not to mention that pulling people forward tends to bend their backs harmfully backward.

The Basics of Torque

Torque is a twisting action that causes an object to rotate similar to the way a force causes an object to translate (move in a linear fashion). A force F applied at a ninety-degree angle to a lever arm or moment arm r creates a torque τ as follows:

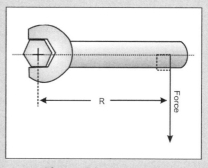

Figure 31: lever arm

$$\tau = F\ (r)$$

The Rules of Rotation/Translation

An object's motion can be categorized as translation, rotation, or some combination of the two. Translation is motion along a smooth path. Rotation is a spinning motion around a pivot point. If the object can move freely, the pivot point will be its center of mass. The center of mass is like a balance point. For example, a child's seesaw will balance if the pivot is located at or under the center of mass.

Like forces, torques can counteract each other. For example, a torque that causes a twisting action in a clockwise direction can be counteracted by one causing a twisting action in the counterclockwise direction. When all the torques are added up, the result is called the net torque; and if the net torque is not zero, the object will have rotational acceleration.

To understand what torques and forces do, let's see what happens when we apply them to a stationary object that could otherwise move freely. Like many things in physics, the resulting motion can be reduced to a few simple rules:

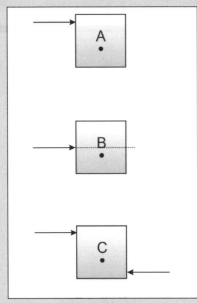

Figure 32: ABC

★ A net force (the result of adding up all the forces) acting on an object will always cause translation in the direction of the force regardless of whether the object rotates or not.

★ A net force that does not act through the object's center of mass or pivot point will cause translation and will create a torque that causes the object to rotate. (Assuming no other torques counteract it. See A.)

★ An object will translate without rotation (see B) if the net force on it acts through the center of mass.

★ An object will rotate without translation (see C) if the net force acting on it is zero but the net torque is not.

Concepts such as center of mass, rotation, and translation are key to understanding many martial arts. Practitioners of Aikido (considered the primary martial art of Steven Segal) actually spend time meditating on their centers of mass in order to remain balanced while sending their opponents flying. Attackers foolish enough to charge headlong at an Aikido master will likely find their linear velocity tuned into head over heels rotation. Up to a point, the same mastery of applied physics can be simulated with clever camera and wire tricks that make even moderately trained actors look like martial artists. If overdone, however, the wire and camera stunts merely make the actors look like cartoon parodies.

KNIFE THROWING

Why would anyone in a life-or-death struggle want to throw away a perfectly good knife? Unlike a gun, a knife never runs out of bullets. Kept in the hand it can be used to attack or defend again and again. When thrown it has to stick in the target to do any real harm, but that's just the first requirement. There's a reason killers often stab their victims multiple times, and it's not just viciousness. Stab wounds, even well-placed ones, are usually not immediately fatal. So, of course, knife throwing is a routine staple of action movies and invariably drops its victims instantly.

The first problem with knife throwing can be summed up in a word: rotation. Typically, a knife thrower will grasp the blade near the tip and raise the knife slightly over his head so that the hilt is pointed slightly backward. As the knife thrower swings his arm downward and forward, the hilt will be pointed toward the target just before release. The knife will have generally rotated at least a fourth of a turn during the act of throwing, even before it leaves the knife thrower's hand. To get this rotation, a knife thrower has to create a force on the knife that does not act through the knife's center of mass. Some knife throwers prefer to grip the hilt rather than the blade during throwing. Either way, the knife will be rotating in what could be called a forward direction as it leaves the hand and flies toward the target.

To stick the knife, the tip has to rotate more or less forward (within ±30°) when the knife strikes its target, or the knife may as well be a stone. Knife-throwing performers generally train themselves to throw their knives so that the blades rotate at a repeatable rate. Throwers then position themselves at a known distance from the target. Using this system along with a lot of practice, good knife throwers can

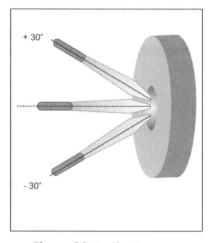

Figure 33: knife throw

stick their blades consistently. Place knife throwers at a different distance, and they lose at least some of their advantage. They can still stick a knife but have to estimate the distance and alter the way they throw enough to make their knives rotate properly.

By holding the knife at the hilt and snapping the wrist correctly as the knife is thrown, it's actually possible to make a knife rotate backward from its normal direction of rotation as it travels toward its target. This would usually be considered a mistake, but done with moderation it can reduce the knife's overall rotation rate.

So-called spear-style or rotation-free knife throwers use heavy knives usually 1 foot (30 cm) or more in length to minimize the tendency to rotate. By adjusting their grip, wrist snap (as mentioned above), and release point, along with lots of practice, they can significantly reduce the rate of rotation. Even then, however, their knives usually rotate at least one-fourth of a turn before striking the target. The stronger the throwing arm and the higher the linear velocity of the knife, the farther it travels before overrotating and failing to stick. Spear-style knife throwing can produce consistent results at distances under 15 feet (4.6 m). By contrast, the Olympic javelin throw record was 321.1 feet (98.49 m) by Jan Železný on May 25, 1996. A well-designed spear thrown by a less talented person would not travel as far but could easily be "stuck" at distances on the order of 100 feet without any worry of rotation.

Spears can be easily thrown and stuck in targets without rotating while knives cannot because the throwing force applied to a spear almost has to act through its center of mass. In addition, there's something called rotational inertia. Inertia is defined as resistance to a change in motion. With linear or translational motion (motion along a smooth curve or line) inertia is equal to an object's mass. Rotational inertia, or the resistance to changes in rotation, is more complex. Like its linear cousin, it is directly proportional to mass but is also related to how the mass is distributed. Different shapes have different rotational inertia values even when their masses are equal. A knife or spear has a shape similar to a rod, which yields a rotational inertia as follows:

$$(\text{rotational inertia}) = 1/3 \, (\text{mass})(\text{length})^2$$

According to the above equation, a six-foot-long spear is going to have at least thirty-six times more rotational inertia than a one-foot-long knife just from length alone. In most cases the spear will also have more mass than the knife, making the spear's rotational inertia even higher. If the force used to throw the spear is slightly misaligned with the center of mass, the spear's rotational inertia will resist the tendency to rotate far better than the knife would in the same situation.

The second problem with knife throwing's effectiveness is bone; bone is hard to penetrate and tends to show up in inconvenient places for the knife thrower—for example, as protective armor for vital organs. A thrown knife delivers a single stab wound (assuming it sticks in the victim). If the knife hits bone, it may not penetrate far enough to inflict substantial damage. If it does penetrate, it may not disable a victim fast enough to prevent him or her from fighting back or calling for help. Even if stabbed directly in the heart, a victim can remain conscious for about 10 seconds. A highly motivated, adrenaline-fueled, or drugged victim can continue to fight back during at least some of this time. Stabbings are not at all like they're depicted in the movies: victims don't just quietly accept their demise the instant they are stabbed.

Although throwing a knife significantly reduces its effectiveness, it does extend its range—by at least a few meters. If a wild-eyed terrorist across the room announces murderous intentions and brandishes a loaded AK-47, then by all means hurl a dagger at him. Why not? It might disrupt his aim. Rush him and hopefully wrestle away his weapon. But, other than desperate situations or circus acts, it's best to hold on to one's blade. Yes, having multiple knives helps, but even then, a thrown knife is more feasible as a distraction than an actual killing technique.

The movie *Cellular* [PGP] (2004) cleverly illustrates how a blade can be used effectively as a weapon. In the movie Jessica Martin (Kim Basinger) is kidnapped and left in a room with a smashed phone. She manages to piece it together well enough to make a highly improbable call to a random stranger's cell phone. He (Chris Evans) spends the rest of the movie trying to rescue her. Not one to helplessly wait for Prince Charming, Basinger— a high school science teacher (not a good person to mess with)— knows anatomy and slices one of her kidnaper's major arteries when he attacks her. As he quickly weakens, she stays just out of range from his grasp long enough for him to bleed to death, a satisfying moment for the audience.

While having a knife incurs a significant advantage over an unarmed opponent, the problem with knife fights is that both opponents have them. At the close range of knife fights, cutting an opponent is easy, but stabbing one in the vitals is hard. On the other hand, the main difference between slicing blood vessels in the arms and legs versus stabbing a vital organ is mostly the amount of time it takes for one's opponent to lose enough blood to pass out or give up. The key to winning a serious knife fight is to make your opponent bleed faster than you do.

FIREARM SPIN

In movies, skill with firearms is as remarkable and common as knife-throwing ability. Shooting from the hip, as discussed in Chapter 2, has long been a Hollywood specialty, but now there is an even cooler handgun technique appearing in movies: the horizontal grip. Hollywood has spun yet another cliché without considering the physics of rotation.

Normally a handgun is gripped with the fist in a vertical position. This enables one to use the gun sights—a handy feature if one wants to hit something. When a handgun is fired in the vertical position, the gun barrel is above the fist and the recoil creates a force that pushes backward above the wrist joint, creating a twisting action on the fist that rotates both it and the gun barrel upward. Conveniently, gravity helps rotate it back, more or less to its original position. This up-and-down rotation can cause aiming errors in the vertical dimension. But the vital area on a person-sized target is about twice as high vertically as it is wide horizontally. If a shooter is going to have an aiming problem, it's best to have it in the vertical dimension.

Torque and Rotation of Firearms

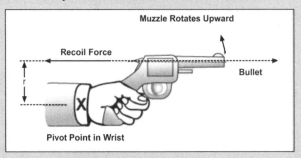

Figure 34: hand holding gun

The barrel of any type handgun will typically be located above the pivot point in the wrist and will generate a torque equal to the recoil force times the distance r, as shown in the diagram. This torque will rotate the gun barrel upward as shown.

Grip a handgun, turn the fist horizontally, and it's oh so cool, but all of the advantages mentioned above disappear. The first to go is the ability to effectively use gun sights. The head is in the wrong position. It's too high. Raise the fist above shoulder level and it's possible to get the sights to a usable level, but it places the arm in an awkward position. The handgun's sights must also be adjusted for shooting with a horizontal grip, or accuracy is going to suffer.

When the shooter fires using a horizontal grip, the recoil will rotate the handgun horizontally and—guess what—there is no horizontal gravity to help restore it to its original position. Resulting errors are going to scatter shots horizontally, where they are more likely to miss a person-sized target. Okay, so maybe Billy Bob out in the Texas Panhandle has mastered the technique after years of practice on rattlesnakes. But it still doesn't change the fact that a horizontal handgun grip is ridiculous for most people. It's a good way to accidentally shoot an innocent bystander.

The confusion about rotation and translation (moving in a linear fashion) doesn't just apply to handgun grip but also to how victims—hopefully bad guys—react when shot. Although conservation of momentum (see Chapter 13) rules out the cliché of shooting victims

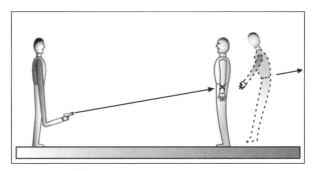

Figure 35: long arm shooting

being blown off their feet and sent flying backward through the nearest glass object, the conditions required for translation also cast doubt on this scenario. To throw a person violently backward and slightly upward, a shotgun or handgun blast (as depicted below) has to hit in line with the person's center of mass (about 2 in. below the navel). This means it has to be fired by a long-armed chimpanzee who can hold the weapon below knee level and fire it in a slightly upward direction. Shooting a person above the belt would tend to rotate him backward. Shooting him in the shoulder would add a spin similar to an ice skater's spin but much more mild. Neither of these effects would send a person flying backward.

ROTATING CARS

Although people can theoretically act in unpredictable ways, inanimate objects such as cars cannot. Yet, even here Hollywood has difficulty. Watch a few hours of reality TV filmed by police-car-dashboard cameras, and it's clear that cars rarely go airborne when they crash. Watch a few hours of Hollywood chase scenes, and cars routinely go airborne while doing more twists than an Olympic platform diver. Even when colliding with the backside of a parked car, the front end of the moving car invariably flips upward, sending it airborne with a half twist that causes it to land on its roof several car lengths in front of the parked vehicle.

To get the entire car airborne, the net force on it has to be large and upward. To flip up the car's front, an upward force has to act on the front to cause it to vertically rotate around the car's center of mass. To twist or rotate the car like a huge slow-moving bullet so that the car ends up on its top, a force has to be applied to one side of the car. Driving a car at high speed up a ramp gives it the large upward force needed to get it airborne. Adding a kicker plate gives the front end an upward rotation. (Kicker plates are short ramps that act only on the front wheels and then fall away.) Using the kicker primarily on one side gives the car the twisting motion that causes it to end up landing on its top. Placing the camera in front of the parked car hides the ramps, kicker plates, and so forth.

Air cannons can also be used to provide missing forces. These are large-diameter, short-barreled devices that are specially mounted inside cars and aimed at the ground. They generally shoot a section of a telephone pole propelled by air pressure. These act like large super-fast hydraulic jacks that almost instantly elevate the vehicle. While it might sound odd to use sections of telephone poles, they are just about the right size and weight, not to mention readily available. Some of the cannons are designed to use black-powder charges when fire and smoke are needed during a crash. If a moving car is supposed to roll over sideways, the cannon is mounted to the right of the driver where a passenger would normally sit. If the moving car is to flip with the car's trunk rotating upward and forward over the cars' hood, the cannon is installed near the center of the car's trunk. By moving the cannon around relative to the car's center of mass, it's possible to get just about any type of rotation desired. When combined with ramps and a high forward speed, the stunt director can choreograph whatever type of motion is needed.

Roaring down the highway, pursued by submachine-gun wielding albino twins (*Matrix II*), Trinity triggered one of the most famous car wrecks in cinematic history. To pull off this scene, some of the stunt cars actually towed ramps behind them as they sped down the road. When the stunt drivers in ramp-towing cars slammed on their brakes, cars behind them continued up the ramps and went airborne over the ramp-towing cars. Various combinations of air cannons and ramps had cars rolling, twisting, and flipping, all at the same time, in every conceivable manner. It was a veritable physics ballet—impressive and entertaining but not realistic.

It's impossible not to admire the technical skill and daring that goes into spectacular car-wreck scenes. The individuals designing them are applied physicists, who must possess a finely tuned understanding of forces. The stunt drivers performing them are risk takers who must put their very lives in danger. Yet movies based entirely on spectacular car wrecks are like restaurants based on food fights. While experiencing them might be fun, they leave one hungry for sustenance.

Summary of Movie Physics Rating Rubrics

The following is a summary of the key points discussed in this chapter that affect a movie's physics quality rating. These are ranked according to the seriousness of the problem. Minuses [—] rank from 1 to 3, 3 being the worst. However, when a movie gets something right that sets it apart, it gets the equivalent of a get-out-of-jail-free card. These are ranked with pluses [+] from 1 to 3, 3 being the best.

[–] Above-the-belt kicks and punches that throw victims horizontally backward significant distances.

[–] Throwing one's knife as anything other than a pastime, circus stunt, or last-ditch tactic. Having multiple knives helps, but even then, a thrown knife is more effective as a distraction than an actual killing technique.

[–] The horizontal handgun grip used at any distance other than point blank.

[–] Spectacular car wrecks that could not happen in reality without ramps and air cannons.

[+] Winning a knife fight in an understated manner by severing an artery other than the carotid.

CHAPTER 17

HOLLYWOOD DISASTERS:
Global Warming, Tsunamis, Tornadoes, and Other Big Winds

GLOBAL EXAGGERATION

It's the ultimate science experiment: dump trillions of tons of CO_2 into the atmosphere for decades and see what happens. It takes a while but then, according to *The Day After Tomorrow* [RP] (2004), snow blankets Bombay, super-sized hail hammers Hong Kong, and terrible twisters trash Los Angeles (F5s on a scale of F1 to F5). Dare to engage in extramarital sex or, worse yet, inane live news coverage as one of these tornadoes approaches, and your demise is certain. But that's just the beginning. The north polar ice cap suddenly melts, shutting down Atlantic currents and in the process triggers a massive hurricane-like blizzard covering North America. Since the North Pole's ice cap floats, melting it does little to raise ocean levels, but then apparently ice sheets in Antarctica and Greenland also suddenly melt causing an immediate change in ocean level. This immediate change sends a super-tsunami selectively sweeping into New York City. Frigid upper layers of the atmosphere plummet downward and freeze people mid-conversation, and then comes the snow—biblical amounts of it.

In reality, the whole state of California—it's definitely not Kansas, Toto—has never recorded a tornado rated greater than an F2[23], but there's no law of physics that says it can't. An ice age in North America from the season's first snowstorm is also far-fetched, but again there's no law against it. Why, however, would New York City be a magnet for super-tsunamis that apparently hit nowhere else? The scientific explanation is obvious: the Statue of Liberty. It washed up on the beach symbolizing the decline of insensitive and thoughtless humans replaced by apes in *Planet of the Apes* (1968). It protruded from a frozen New York Harbor in *A.I.* (2001), symbolizing the decline of insensitive and thoughtless humans replaced by skinny robots. So, what could be better for symbolizing the decline of insensitive and thoughtless fossil-fuel-burning fools? Have the old gal get slapped in the kisser by a global-warming-induced super tsunami; that should do it. But destroy her—never! It would destroy all hope of a happy ending. So the movie leaves her protruding from the frozen harbor, still standing, as a symbol of hope.

At 305 feet (93 m) tall (including the pedestal)[24], the venerable lady also puts things in perspective. Judging from its level on the statue, the tsunami has to be around 240 feet (72.8 m) high. This height means that the statue's feet will be under nearly 90 feet (27.4 m) of water. Unless the statue instantly fills with water, its feet will be subject to about 2.7 atmospheres of water pressure, enough to crush the statue's 0.09375-inch-thick (0.237 cm) copper sheet metal walls. Keep in mind that each square foot of the thin sheet metal in the base of the statue would have to resist a total force of over 5,700 pounds (25,000 N). Add the massive impact of the wave, and the grand lady's days of symbolizing are over. She's going to be a twisted sunken wreck.

When it hits the shore, the giant 240-foot- (72.8-m-) high wave would again be far more devastating than depicted. By comparison the Indian Ocean tsunami on December 26, 2004, was as much as 115 feet (35 m) high as it rolled ashore in Sumatra near the epicenter of the earthquake that caused it, killing about 320,000 people. In Thailand the tsunami was down to at most 34.7 feet (10.6 m) high with a death total of 14,000 people. By the time it reached India, the tsunami was less than 8 feet (2.4 m) high—a mere shadow of its younger self—yet still caused about 22,000 deaths. Overall, more than 400,000 people died as a result of the tsunami's devastating power. A 240-foot- (72.8-m-) high super-tsunami hitting New York City is going to make the December 2004 tsunami look mild in terms of death and destruction.

Static water pressure would be 7.25 atmospheres (atm) at the bottom of a 240-foot wave as compared to an estimated pressure of 4.1 atmospheres under the Hiroshima atomic bomb blast. Like the Statue of Liberty, a building would be unaffected by the static water

SUPER-TSUNAMI MATHEMATICS

Mathematically predicting the destructive potential of the super-tsunami would be extremely complex. However, we can get a rough idea of its destructive potential by comparing it to a wind with similar kinetic energy. The famous Bernoulli equation indicates that when a moving fluid such as a wind or water flow is stopped, say, by running into a wall, the kinetic energy of the flowing fluid will be converted into pressure acting on the wall. Obviously, if the pressure is too high the wall collapses.

To approximate how the pressure created by flowing water compares with wind, we can use the kinetic energy term from Bernouli's equation and solve for the velocity of air as follows:

$$\tfrac{1}{2} \text{ (DENSITY SEA WATER)(VELOCITY WATER)}^2 =$$
$$\tfrac{1}{2} \text{ (DENSITY AIR)(VELOCITY AIR)}^2$$

Solving for air velocity yields:

$$\textbf{(VELOCITY AIR) = [(DENSITY SEA WATER) /}$$
$$\textbf{(DENSITY AIR)]}\tfrac{1}{2} \textbf{ (VELOCITY WATER)}$$

$$= [(1{,}027 \text{ kg/m}^3) / (1.25 \text{ kg/m}^3)]\tfrac{1}{2} \text{ (velocity water)}$$
$$= 29 \text{ (velocity water)}$$

Using the above relationship, a sedate water velocity of ten miles per hour will be equal to an air velocity of 290 miles per hour from a kinetic energy standpoint—the equivalent of the wind speed in a category F5 tornado. Ramp the water velocity up to thirty-five miles per hour and the available kinetic energy would be equivalent to a 1,000 miles per hour of wind, similar to the wind speed under the Hiroshima bomb blast, estimated at 980 miles per hour (1,580 kph)[26]. Since anyone attempting to measure it would likely be killed, good data for the velocity of a tsunami after coming ashore is hard to find; however, it's typically estimated to be between ten and thirty-five miles per hour. A 240-foot-high wave would probably be faster.

pressure if the interior and exterior water levels rose at the same rate, but that's unlikely. The first wall to be hit by the wave would be instantaneously submerged, while the wall on the opposite side of the building remained dry. This condition, however, would not last long. A wave traveling thirty-five miles per hour (56 kph) will cover a distance of 51 feet (15.5 m) in a second. With the exterior water level rising around the building's entire perimeter in a matter of seconds, the resulting exterior water pressure would be more than enough to simultaneously implode windows and sections of walls. The water gushing in would act like a giant piston compressing air in the building into a high-pressure shock wave that would travel upward, blowing out windows and sections of wall as it progressed. It would be followed by a water hammer gushing upward though every possible path. The resulting structural damage could bring the building down, but that's only part of the picture.

From a kinetic energy standpoint, a wall of water traveling 35 miles per hour would be equivalent to a 1,000-mile-per-hour (1600 kph) wind! Since the wind speed directly beneath the Hiroshima bomb blast was estimated to be 980 miles per hour (1,580 kph), it's clear that a 240-foot-high super-tsunami would be horrendously destructive. By comparison, the highest tornado wind speed ever measured was 318 miles per hour (512 kph). Wind pressures act on almost the entire side of a building from top to bottom, while wave action will affect mostly the lower part, so comparisons of relative kinetic energy don't necessarily correlate with damage. However, weakening a building's support at its base is an invitation for collapse.

The combination of static pressure from the depth of water and dynamic pressure from the wave's motion would likely cause structural failure in the affected high-rise buildings with the possibility of a domino-like collapse. Even if a building did not collapse into one of its neighbors, the collapse would cause a local tsunami as falling building materials slammed into the water. Few structures would be capable of withstanding the abuse of being hit by a super-tsunami followed by the wave action of local mini-tsunamis. New York City would largely lie in ruins.

Why Buildings Collapse

The collapse of the twin towers on 9-11 due to airplane impact and resulting fires, along with the destruction of the Alfred P. Murrah Federal Building in Oklahoma City from a truck bomb, have demonstrated to most people that modern buildings are not as indestructible as they seem. Yet conspiracy buffs seem convinced that only explosives planted on the inside can bring a modern building down. For the Murrah Building, conspiracy theories generally go something like this:

The buildings were constructed of such and such materials rated for thousands of pounds per square inch, how could an explosion outside the building with far less pounds per square inch demolish it? Obviously, explosives were planted inside.

Simply put, the "thousands of pounds per square inch" refers to the maximum internal stress the material can handle before failing catastrophically, not the maximum external pressure, pressure that might be caused by a bomb blast. Stress and pressure are very different quantities, although they use the same units (psi). Stresses are internal and depend on an object's material and shape as well as the load applied to it. Pressure is a type of external load. Calculating stresses is too involved to describe in anything less than a text book, but suffice it to say that the internal stresses on the structural parts of buildings are usually far higher than the external pressures on them. For example, the maximum wind pressure loads buildings are designed to withstand are generally less than 0.74 pounds per square inch (0.05 atm). Although this load sounds miniscule, such a pressure would have created a sideways force of about 270,000 pounds on the vertical wall of each floor in the Murrah Building for a total force of over 2.4 million pounds on the building. To produce such pressures and forces with natural phenomenon would require something like a record-breaking hurricane or a direct hit by a tornado. The Oklahoma bomb blast pressure was easily higher.

Elastic stability or buckling resistance of the columns support-ing a building's load is also a factor in collapse. Cut a soda straw about 3 inches long and compress it like a column running between your index finger and thumb. The straw will withstand a considerable force before it buckles. Repeat the experiment, only this time push lightly sideways on the straw as you com-press it. The straw will quickly buckle. Sideways forces applied to support columns by explosions or walls of water are a very bad idea if the goal is to keep a building standing.

Similar conspiracy theories—multiple bombs planted by insiders—exist for the twin towers. According to conspiracy theorists, the fire inside the towers could not have melted the steel structure that supported the top floors, thereby leading to their collapse, and here the theorists are right. The steel support beams would not have melted, because they didn't have to. At around 800 degrees Fahrenheit (430°C) steel beams or trusses begin to significantly lose both their strength and stiffness.

In the twin towers most of the supporting steel columns were located in the outer walls and were connected together by long, light-weight steel trusses in the floors. With fire temperatures inside the towers easily exceeding 800 degrees Fahrenheit, the floor trusses sagged. Stiff trusses normally create little sideways force on the steel columns in the outer walls. But let the trusses sag, and they become more cable-like and capable of creating sideways forces many times the downward weight of forces acting on them. As a result of these forces, the outer wall's columns were pulled inward until they snapped, sending the whole top of each tower careening down on the floors below in a chain-reaction collapse.

Conspiracy buffs counter by claiming that no modern building has collapsed due to fire. Maybe it has something to do with the fact that, except for the twin towers, no modern building has ever been hit at full speed by a fuel-laden jumbo jet that destroyed a significant part of the building's support structure and caused parts of the floors in the impact zone to collapse on those below, not to mention leaving the weight of many jumbo jet parts scattered on the floor in the burning inferno. From the beginning, the floors and structures in the impact area were overloaded.

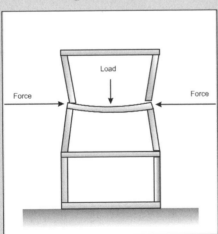

Figure 36: force/load

The conspiracy buffs continue by asking how a few collapsing floors could bring down an entire building. Of course, they fail to note that the "few" collapsing floors (thirty-three floors for tower two, and seventeen for tower one) would have been tall enough in themselves to be major buildings in many communities.

Floors are usually designed for higher-pressure loads (not including the weight of the floor itself) than walls, but the maximum pressure is still generally less than 0.05 atmospheres. Pile the concrete and steel debris from several collapsed floors atop a good one, and its pressure rating is easily exceeded. But, in the twin tower collapse, the chunks of debris were not just piled, they fell with considerable impact. The result was an almost instant failure of the floor being struck.

The floors might have simply pancaked to the bottom, leaving the walls standing like a huge hollow tube, except that the floors were firmly attached to the walls. Overloading a floor enough to make it sag and collapse would have produced a substantial sideways force on the walls, pulling them inward, similar to the effect of making the floors sag from high temperature only on a much faster time scale. The result: structural steel in the walls would have buckled and snapped as the floors failed.

Conspiracy theorists continue by asking how a building collapse could pulverize the structure's concrete and give the appearance of a blast wave emanating outward. The answer is straightforward: Energy used to lift building materials over a construction period lasting years was stored in the buildings as potential energy—an amount roughly equal to 500,000 pounds of TNT (230,000 kg TNT) per tower, easily enough to pulverize much of the building's concrete. When each tower collapsed, all the energy was released. As top floors fell they acted like a gigantic piston compressing the air in the floors below. This compressed air had no way to escape except by blasting outward and downward, taking anything loose with it, such as concrete dust. By the time the building approached the ground, the velocity of the escaping air was high enough to throw hapless victims violently sideways. Placing a few thousand pounds of explosives inside the building would have made little difference.

Since the super-tsunami in *The Day After Tomorrow* does relatively little damage to the Statue of Liberty, the statue once again provides a useful reference point when the water eventually recedes. Allowing for about 20 feet (6.07 m) of snow, the water level had to remain over 150 feet (45.5 m) higher than normal after the tsunami had receded. To raise ocean levels by 150 feet (45.5 m), about 75 percent of Antarctica's ice would have to melt. That would take about 2.6 years, assuming that all solar energy available to Earth went entirely into melting Antarctica's ice and that the ice was already warmed up to 0 degrees Celsius[26] Obviously, this is only a fraction of the time required for melting, and so a super-tsunami with an immediate 150-foot increase in ocean level caused by melting ice is absurd.

The super-tsunami depicted in the movie could have been a storm surge brought on by high winds, but that's also ridiculous. At 240 feet (72.8 m) high, the super-tsunami is about 215 feet (65.2 m) higher than the unusually high storm surge that occurred in Louisiana during hurricane Camille (1969). That surge was caused by maximum wind speeds near 200 miles per hour (322 kph)[27]. A 240-foot- (72.8-m-) high storm surge would be virtually impossible without help from a catastrophic event such as an asteroid strike or massive land slide. Even then it's unlikely that New York City would be inundated while Washington, D.C., was left untouched (as depicted in the movie).

On the one hand, the giant wave produced far less damage in the movie than would have been experienced in reality had such a wave existed. On the other hand, there were no conditions in the movie that could have caused the wave in the first place. It's classical Hollywood logic: two conflicting mistakes equal perfection. Serious as it is, global-warming effects are not going to make the same dramatic entrance on the world's stage as they did in *The Day After Tomorrow*. The behemoth global-warming-induced wave is nothing more than a global-sized exaggeration; the potential long-term dangers of global warming are not.

THE ULTIMATE ANTI-CHAOS THEORY

The biggest surprise in *The Day After Tomorrow* is the total lack of reference to chaos theory. The theory had its roots in meteorology, and while far too rich to explain in a few paragraphs it can be partly understood by defining the "butterfly effect," a term coined in the 1960s by Edward Lorenz[28]. According to Lorenz, chaotic systems such as the weather are extremely sensitive to initial conditions, and even slight changes in initial conditions can radically change such a system's behavior over time. Weather is so sensitive that, theoretically, air currents caused by a butterfly flapping its wings in Beijing could eventually cause a tornado in Kansas. Imagine what a radical change such as global warming could do.

It's not as though Hollywood is unaware of chaos theory—in *Jurassic Park* [PGP-13] (1993) Malcolm (Jeff Goldblum) babbles endlessly about it. Although his arguments were muddled, they indicated that dinosaurs brought back to life in the movie would eventually become able to reproduce in spite of precautions to the contrary. Sure enough they did, but big deal. If humanity could wipe out everything from the woolly mammoth to the carrier pigeon, then surely a few hundred dinosaurs couldn't be all that hard to annihilate. It's not like they would be hard to find.

Although *The Day After Tomorrow* does offer some explanation for how global warming can trigger an ice age, the paradox doesn't make sense at a gut level. That's not to say the concept is wrong, but rather an ice age would have seemed more plausible in the context of chaos theory. In that setting, unexpected or even opposite results seem reasonable. But referring to chaos theory would have made the magical weather predictions of the movie's climatologist hero look silly. Weather accuracy is very sensitive to data available at the time a weather forecast is made—the greater the amount and accuracy of data, the greater the accuracy of the prediction, but only up to a point. Lorenz claimed that weather predictions could never be accurately made more than about two weeks ahead because that was the limit of humanity's ability to collect and analyze the required data. Extending beyond that time span would quickly require a forecaster to know the condition of every atom in the universe, he claimed. Nevertheless, the movie's climatologist nailed his forecast of a deadly ice-age-producing storm days ahead of time by using a computer simulation based on a few ice samples from Antarctica—what a guy.

With such a dire prediction in hand, what can be done? Why, of course, evacuate the entire southern United States to Mexico—northerners are doomed, but they may actually be the lucky ones. Imagine this joy: being trapped for endless days in a traffic-snarl hundreds of miles long during a major blizzard. What a great problem-solving idea: the ultimate antichaos theory—death.

Summary of Movie Physics Rating Rubrics

The following is a summary of the key points discussed in this chapter that affect a movie's physics quality rating. These are ranked according to the seriousness of the problem. Minuses [–] rank from 1 to 3, 3 being the worst. However, when a movie gets something right that sets it apart, it gets the equivalent of a get-out-of-jail-free card. These are ranked with pluses [+] from 1 to 3, 3 being the best.

[–] [–] Highly exaggerated storm conditions.

[–] [–] Seriously underestimating the damage that would be caused by highly exaggerated storm conditions.

[–] [–] Seriously overestimating or underestimating damage done to structures by natural phenomena or acts of terrorism. (Movies should use computer simulations and engineering studies to make these estimations.)

[–] Incredible computer-based predictions of chaotic systems, such as weather systems or the stock market, made in ridiculously short periods of time and based on flimsy amounts of data.

[–] Scientists spouting dramatic and authoritative-sounding solutions or pronouncements that make no sense from a logical or scientific perspective (such as evacuating vast territories in the middle of a blizzard).

[–] Evacuating vast territories in the midst of foul weather. (Please, didn't hurricane Katrina teach us anything?)

CHAPTER 18

THE MOVIEMAKER'S COOKBOOK:
Cigarettes as Lighters, Exploding Cars, Burning Bugs, and Other Recipes for Foolishness

CIGARETTES AS LIGHTERS

Call them cancer sticks, smokes, or whatever, but by any name, cigarettes are an amazing piece of technology—deliberately designed for reliable failure. Obviously combustible, they fail to burst into flame, even when burning at temperatures in excess of 1,000 degrees Fahrenheit (540°C)[29]. By design they smolder. Pull air through them, and they smolder really brightly and really quickly, but that's all they do. They just don't burst into flame.

Light the wrong end (the filter), and it burns like a candle with a bright yellow-orange flame. But let a cigarette burn down to the filter and, remarkably, the tobacco smoldering at over 1,000 degrees Fahrenheit does nothing. It reliably fails to light the filter's combustible cellulose acetate fibers.

So what happens when a movie villain flicks his cigarette into a puddle of gasoline for some nefarious purpose? It bursts into flame. Try the same experiment in the real world, and the gasoline will snuff out the cigarette as soon as it lands in the liquid. Okay, it's vapor, not liquid, that burns. The cigarette, however, has to pass through a combustible mixture of gasoline vapor and air before it lands in the puddle. Even if a smoker draws gasoline-vapor laced air through it, a cigarette will still not reliably ignite the vapor.

It's Not So Simple—The Fire Triangle

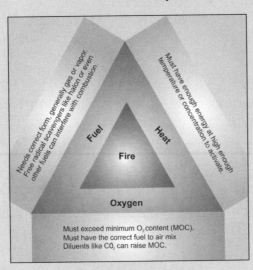

Just about every grade school kid learns about the fire triangle and the three requirements for creating or putting out a fire. Although the triangle is a good starting point, actual combustion is not so simple. The three conditions needed for fire have to meet a number of conditions of their own. This is why it's difficult (if not impossible) to ignite gasoline with either bullets (especially handgun bullets) or cigarettes.

Figure 37: fire diagram

Combustion is a multistep process that first depends on the formation of free radicals. Free radicals are atoms or molecules with one or more unpaired electrons that provide a reaction site. After formation, the right free radicals have to bump into each other at the right moment in the right way—by a random process—so that they can bond and release energy. For example, an O_2 molecule must absorb enough energy to break the double bond holding the two oxygen atoms together in order to form two free radicals—one of many types that must form. In turn, these O free radicals must bump into a carbon free radical to form CO or into a CO free radical to form CO_2 or an H to form OH or any number of other free radicals to form other interim compounds. In the end, a hydrocarbon fuel (made of Cs and Hs) becomes CO_2 and H_2O—that is, if enough activation energy is present to start the process and enough excess oxygen to sustain it.

When heat is liberated during combustion, it must provide the energy to continue activating the various steps in combustion, or the fire will go out. Simply diluting the reactants so that heat is transferred to nonreacting molecules instead of reacting ones can be enough to stop a combustion process. Interfering with the free radicals formed in interim steps can also end a combustion process. When heated, halogenated compounds such as Halon 1211 (CF_2BrCl) form their own free radicals that effectively mop up the free radicals in combustion processes and extinguish fires. Generally, Halon is effective in concentrations of only 5 percent to 10 percent. Unfortunately, such compounds are also highly effective at destroying the ozone layer and have pretty much been banned. All in all, starting and maintaining a fire is not as simple as it looks.

Unconvinced by testimony from several sources, a group of staff and volunteers at Intuitor.com tested the gasoline-igniting ability of cigarettes over three hundred times without a single success. They tried tossing lit cigarettes into a small puddle of gasoline, holding them (with really long tongs) in the vapor-air mixture at various distances above the puddle, placing up to forty smoldering cigarettes in and around the puddle, and using gasoline-soaked paper towels instead of a puddle, all with no effect. They devised a smoking machine and smoked numerous cigarettes directly over the puddle. Using the machine, they blew air backward through smoldering cigarettes, making sparks fly off the brightly glowing tip into the gasoline—but still no ignition. They lit an unfiltered cigarette and placed the unlit end in the gasoline. The flammable fluid wicked its way to the smoldering end and snuffed it out.

In distress, the experimenters began thinking the gasoline was defective. They held a match above the puddle (at the end of a long pole) and—poof—the fuel burst into flame. They repeated this process several times, always with the same result. Clearly, cigarettes were just not reliable igniters even though matches were.

The myth of cigarettes reliably igniting gasoline pervades popular culture so deeply it has worked its way into novels. In *The Partner*, John Grisham has a character splash gasoline around his vehicle (ironically, a Chevy Blazer), light a cigarette, and throw it in the gas, igniting an inferno. Tell a group of friends that the lit cigarette wouldn't have ignited the gasoline, and invariably at least one will offer, "but they do it in movies." (A few words of advice for aspiring novelists: don't use movies for source material. The real-life experiences touted by writing teachers are actually much better.)

So, how does a cigarette ignite a gasoline puddle in a movie? It doesn't. At the right moment, a special-effects technician pushes a button that fires an electronic igniter located appropriately near the fuel. The camera is placed so the igniter can't be seen. Sometimes, the fuel is lit by a trick cigarette containing combustible compounds other than tobacco. It is never, however, lit with anything as unreliable as an ordinary cigarette.

The explosive or flammable range for gasoline is only about 1.4 percent to 7.6 percent gasoline vapor in air (21 percent oxygen). Concentrations below 1.4 percent don't give off enough heat to sustain the chain reaction of combustion. Concentrations above 7.6 percent don't have enough oxygen to sustain combustion. Outside these very narrow limits, gasoline cannot be ignited.

Nevertheless, even the right mixture of gasoline vapor and air needs some activation energy for igniting, similar to the way some pressure is needed for tripping a mousetrap. At 536 degrees Fahrenheit (280°C)[30] a combustible gasoline-air mixture already has enough internal energy to trigger combustion with no further energy input. Having a gasoline-air mixture contact an object at 536 degrees Fahrenheit (280°C) or higher, however, does not automatically heat the mixture to 536 degrees Fahrenheit (280°C). Heat must be transferred from the high-temperature object into the vapor-air mixture to elevate it to 536 degrees Fahrenheit before ignition occurs. Heat the gasoline-air mixture outside the cigarette and as soon as it gets warm, the vapor-air mixture will rise and mix with colder air, preventing it from reaching ignition temperature. A similar effect prevents firewalkers from burning their feet as they walk across glowing coals. If their feet contact the hot coals briefly, the heat transfer is too slow to cause burns, but stand on the coals too long and the feet get fried.

Is there any other example of a combustion process that will not ignite a combustible fuel-air mixture? Yes, the Davy safety lamp. In the 1800s coal mines could only be illuminated with some sort of flame—better for igniting the mine's methane gas than for illuminating the mine's dark tunnels. The Davy lamp solved this problem by surrounding the flame with a fine wire screen. A flammable methane and air mixture could enter the lamp and make it burn brighter, but the flame would not propagate outward through the screen. Presumably, the screen disrupted heat transfer from the flame to the methane-air mixture and prevented it from reaching its ignition temperature.

Can a cigarette ever light gasoline vapors? Yes, under special conditions, and it makes a convincing fire-safety demonstration in the process. A firefighter sets an object that looks like a skinny chrome-plated metal vase on the table. It has a square base with an 18-inch-long (45.5 cm) vertical tube, about 2 inches (5 cm) in diameter, welded to the base and open on the other end. He fills the tube with pure oxygen (not air), adds a few drops of gasoline, heats the outside slightly to vaporize the fuel, drops a lit cigarette in the open end, and—BOOM!—the gasoline explodes. So, yes, cigarettes can ignite gasoline vapors; yes, one should never risk smoking near gasoline; and yes, smoking is dangerous. But then there aren't many gas stations, warehouses, or fuel dumps with pure oxygen environments.

If anything, the exploding vapor demonstration shows that altering oxygen concentration makes a big difference in how gasoline vapors do or don't ignite. Lower the oxygen content enough to the minimum oxygen concentration (MOC, about 12 percent for gasoline), and the gasoline vapor cannot ignite regardless of its concentration.

If there's a chance of an explosive mixture forming in the empty space of a storage tank, the chemical company that owns it will keep the empty space below the MOC of the tank's contents, often by diluting the empty space's air with nitrogen.

Cigarettes are designed to deliberately create an oxygen-poor, fuel (tobacco) rich environment in order to produce smoke. Draw a combustible mixture of gasoline and air into the smoldering tip, and the burning paper, which has a combustion temperature of 451 degrees Fahrenheit (233°C), along with the tobacco will greedily gobble up most of the available oxygen, lowering oxygen content below gasoline's MOC of 12 percent before the gasoline vapors get to their ignition temperature of 536 degrees Fahrenheit (280°C). What happens to the oxygen? It mostly becomes CO_2, the same stuff that's in CO_2 fire extinguishers. Diluting air's oxygen content with CO_2 is even more effective for fire suppression than diluting it with nitrogen.

Past the glowing tip, fresh air will indeed seep in through the cigarette's porous sides—boosting the oxygen content back up to around 13 percent coming out of the filter[31]. But in the process, it also lowers the temperature well below the gasoline vapor ignition point.

Neither theory nor experiments can totally rule out the possibility of igniting a fire with an oddball cigarette combined with a special set of weather conditions and an unusual blend of gasoline. But there's certainly enough theoretical and experimental information to show that cigarettes won't reliably ignite gasoline—probably due to a combination of poor heat transfer and low oxygen content in the burning tip. What's more, moviemakers know this. So, naturally, it makes perfect sense for them to continue filming clichéd scenes falsely depicting it.

CARS AS BOMBS

Movie bullets can be as effective as cinematic cigarettes for setting off gasoline. Shoot a car's gas tank, and it explodes in a fireball. But how? It's actually not as easy as it looks. First, automobile gas tanks aren't put in exposed locations. They're usually sandwiched between heavy frame members and surrounded by one or more layers of sheet metal in the car's body. The tanks are hard to see, let alone to shoot. Second, if there's any gas in the tank, gasoline is so volatile that enough of it would vaporize to ensure that the fuel-to-air ratio in the tank was too high for combustion.

Stop a bullet suddenly, and it gets really hot because its kinetic energy changes into heat. In theory, a high-powered rifle bullet can reach high enough temperatures to set off a combustible mixture of gasoline vapor and air. But it's just about impossible to stop a bullet abruptly enough when shooting a gas tank. The gas tank doesn't offer enough resistance. The bullet will likely plow right through. If it is stopped inside, it's because it has already lost most of its kinetic energy by penetrating the car's body, frame, or gas-tank wall and will likely not have enough kinetic energy left to ignite anything.

Sometimes, a bullet can cause a spark, especially if the bullet has steel parts or a static electric charge on it, but these are not the bright fiery flashes of light caused by overuse of pyrotechnic compounds in bullet impact special effects. They are modest sparks that are neither effective at starting fires nor clearly visible in normal daylight. For one thing, the sparks will likely occur outside the gas tank—where there are no combustible vapors—when the bullet first contacts the gas tank's wall. For another, most bullets are made of lead or copper-jacketed lead, both materials that can be ground with high-speed grinders without producing any visible sparks, as opposed to grinding steel, which produces a shower of sparks.

Sparks produced by grinding steel are generally orange-yellow in color, indicating a temperature of over 1,832 degrees Fahrenheit (1,000°C), way above the autoignition temperature of gasoline vapor (280°C). Will they ignite gasoline? Typically, no. Again, the doubters at Intuitor.com tested the gasoline-igniting ability of sparks from grinding steel by repeatedly grinding over a small pan of gasoline. The sparks showered into the pan for several minutes with no ignition. Again, to prove that a combustible mixture existed over the gasoline, a lighted match (on the end of a long pole) was held above it, and the mixture burst into flame. Although the sparks themselves were well above the autoignition temperature of a combustible gasoline vapor and air mixture, they did not contain the right amount of thermal energy to heat the combustible mixture around them to a high enough temperature for ignition.

Once again it should be emphasized that neither theory nor experiments can totally rule out the possibility of igniting gasoline with sparks from a grinding operation, and so grinding near a puddle of gasoline has to be considered a very unsafe activity. However, sparks from a normal grinding operation are not a reliable source of ignition. This casts doubt on whether sparks from a steel bullet would be any more effective.

If a vehicle's gas tank is shot with a machine gun and gasoline leaks out, and it forms a combustible mixture, and a bullet just happens to hit in such a way that it makes a spark with just the right amount of energy inside the combustible mixture, then—poof! But that's not likely. Even heavy .50-caliber machine guns in WWII airplanes were found to be unreliable for starting fires of this type.

Incendiary bullets were invented to remedy the problem. They contain a pyrophoric material, which bursts into flame when it hits an object and burns like white phosphorus. Although the fire is very brief, it's also very hot and highly effective at igniting flammable materials. Tracer bullets can also sometimes light fires but are less effective. The tracer compound is designed to produce a streak of light as the bullet flies toward its target so that gunners can see where their shots are going. The tracer compound does not burn as intensely as incendiary material, and it is often largely or entirely used up by the time the bullet reaches its target.

Incendiary bullets are the only type capable of creating the large bright flashes of light on impact commonly portrayed in movies. Outside the military, virtually no one uses incendiary bullets in gunfights. The overdone bullet impact flashes in movies can be traced to *Raiders of the Lost Ark* [PGP-13] (1981). Bullet impact pyrotechnics were overused in this movie to create excitement. The bright flashes certainly give the impression that bullets would be effective fire starters. The movie was a hit and—as is Hollywood's habit—was widely emulated. But that doesn't change the fact that, on impact, real bullets don't make such fiery flashes and very rarely start fires.

Movie car crashes also routinely trigger infernos. Drive a car off a cliff, and it's certain to burst into flame, sometimes even before it hits the ground. A character can flip, roll, and smash his car into the shape of an accordion, and yet walk away without so much as a broken nose. But if it's time to kill him off, he's going to get the fireball. Sometimes for dramatic effect he's granted a few moments to ponder his fate while trapped upside down in the car as a puddle of gasoline ominously oozes beneath him. Nevertheless, a fiery end is assured—sometimes at the hands of a villain who sneers as he tosses a cigarette in the puddle.

Yes, cars can catch on fire and sometimes even explode, but it's rare. The narrow flammable limits of gasoline vapor in air make it unlikely. Not only does the gas tank have to rupture—hardly preordained—but a source of ignition has to occur in combination with an explosive mixture of gasoline and air.

BURNING BUGS

After crash landing on Mars, finding their habitat destroyed, and running out of air, the crew in *Red Planet* [RP] (2000) runs into a locust-like swarm of flesh-eating nematodes that can bore right through their space suits. But it's not all bad news. The nematodes made the oxygen atmosphere, which kept the crew alive when they removed their helmets as their air tanks ran dry. The voracious little guys are also highly flammable and, therefore, easily exterminated. Light one on fire and it bursts into flame, setting off a chain reaction with its nearby cousins in a veritable fireworks display.

Apparently, viewers are supposed to believe that the nasty little critters not only make oxygen from Mars's otherwise CO_2

atmosphere, but also store it in their bodies under high pressure. According to Robert Zubrin, if Mars were terriformed so that it had a breathable atmosphere, the pressure would only be about five pounds per square inch (1/3 atm). Although it would be nearly pure oxygen—a condition favoring rapid combustion—the partial pressure of oxygen would be about the same as on Earth, a condition favoring normal combustion rates. Most of the oxygen needed for making the nematodes burn like fireworks would have to be stored inside them. The oxygen concentration outside the little guys would not be high enough for such spectacular combustion.

It takes at least 3.45 grams of oxygen to completely burn 1 gram of paraffin wax (assuming a composition of $C_{30}H_{62}$). Oxygen free radicals have to "bump into" free radicals from fuel molecules for combustion to occur. As oxygen is consumed, these collisions become less and less likely. So, in reality, significantly more than 3.45 grams of oxygen must be available to insure complete combustion.

Assuming that burning nematode carcasses consume as much oxygen as candle wax, the little guys would need to compress and store large quantities of oxygen at high pressure to be so flammable. The stored oxygen would weigh more than 3.45 times the critter's deflated body weight. If the voracious little guys gave off oxygen as a waste product—enough to create an entire oxygen atmosphere—then why would they want to store and tote around such a heavy load of it?

Producing oxygen from CO_2 is a very expensive process from an energy standpoint, but plants are clearly willing to pay. The process is about the only way they can get the carbon they need for growth. Since plants can't move around, they have to let the carbon come to them as CO_2 carried by the wind. Plants have lots of surface area in their leaves for collecting solar energy, and so it's no problem for them to collect the payment for their energy bills. Small mobile critters such as the nematodes have neither the means nor the need for withdrawing carbon from air or energy from sunlight. They can get all the carbon and energy they need from munching on plants. So, naturally, it makes all kinds of sense for them to waste energy creating an oxygen atmosphere for assisting possible competitors—such altruism.

Ironically, the nematode scenes required that moviegoers remember enough science to know things burn better in pure oxygen. The scenes then sought to kill all further scientific reasoning with the yuck factor—seeing little critters with the charm of cockroaches crawl out of a dead person's nose—along with the shock effect of seeing a burning critter cause a chain reaction of flaming nematodes shooting from random holes in a corpse. After that combination, who's going to notice that the science is silly?

By the end of *Red Planet*, the crew has mostly been eaten, murdered, or otherwise destroyed; the surface of Mars incinerated; the years of effort to establish oxygenating plant growth wiped out; and the hope of a new habitat for people gone. Clearly, the required happy ending is at risk, not to mention the future of humanity. But fear not! Having survived the ordeals, the two remaining crewmembers (a man and a woman) are headed back to Earth. They have not only found romance but a solution to Earth's critical overcrowding and pollution problems: bring a flesh-eating, oxygen-producing nematode back to Earth—certainly the cure for overcrowding if not for pollution.

Summary of Movie Physics Rating Rubrics

The following is a summary of the key points discussed in this chapter that affect a movie's physics quality rating. These are ranked according to the seriousness of the problem. Minuses [–] rank from 1 to 3, 3 being the worst. However, when a movie gets something right that sets it apart, it gets the equivalent of a get-out-of-jail-free card. These are ranked with pluses [+] from 1 to 3, 3 being the best.

[–] [–] Spectacular combustion scenes, such as burning nematodes, that have no logical or reasonable scientific basis.

[–] [–] Contrived happy endings with no logical or reasonable scientific basis.

[–] [–] Lighting gasoline puddles by tossing in lit cigarettes.

[–] [–] Blowing up cars by shooting the gas tank, especially when it's done with a handgun bullet.

[–] Clichéd brightly flashing bullet impacts.

[–] Clichéd fiery car crashes.

CHAPTER 19

WARS VERSUS TREK:
Forgiving versus Forgetting

FORGIVENESS

Interstellar space travel even remotely similar to anything in *Star Trek* or *Star Wars* would require a much deeper understanding of physics—possibly even the discovery of as-yet-unknown principles of physics—not to mention far more advanced engineering capability. Such travel makes for great stories but may be completely impossible. Certainly, with the existing knowledge of physics, the speed and energy required for these space flights are too high and the length of a human lifetime too short (see Chapters 7 and 10). Any movie in the space-travel genre automatically ventures into the hostile depths of insultingly stupid movie physics (ISMP).

Venturing into such depths is not, however, automatically fatal; it's like scuba diving—take the right gear, respect the dangers, and survive. Find some sunken treasure (for a movie: a good story line) and prosper. Under the right conditions, even the depths of ISMP are capable of forgiveness, but there must be reasons for it—good reasons.

GOING WHERE NO MAN HAS GONE

Star Trek has been plagued by plenty of half-correct science and mumbo-jumbo explanations, not to mention inconsistencies, yet is not just redeemable but also groundbreaking, at least for movies and television. To understand its significance requires a temporary digression from movies to TV in order to look at the conditions when *Star Trek* first appeared. The TV series that spawned the *Star Trek* movies made a an inventive shift from movie-TV tradition by featuring large unaerodynamic spaceships, built in outer space and designed to stay there, thereby eliminating the enormous fuel resources needed to lift such ships off the surface. By never entering the atmosphere, the ships did not need to be aerodynamic. The large ship size stemmed from the large travel times and distances of interstellar space travel, requiring enormous amounts of resources.

Star Trek spacecraft again broke with movie-TV tradition by not using conventional rocket thrusters for propulsion. The starship Enterprise's impulse engines are indeed thrusters but are supposedly powered by a fusion process similar to that used in the hydrogen bomb, albeit in a much more controlled manner. The ship's warp drives were also a movie-TV innovation, barely conceivable within the known boundaries of physics. Still, the drives were created in recognition of the fact that conventional rocket thrusters—even upgraded fusion-powered devices—could not possibly provide the incredibly high speeds needed for interstellar space travel.

Star Trek broke with movie-TV tradition yet again by fueling its spacecraft with the ultimate energy source: antimatter. True, using such a fuel is extremely unlikely but, at least, conceivable.

It is also about the only conceivable energy source condensed enough to power an enormous spacecraft on lengthy interstellar space journeys.

Some fictional inventions, such as inertial dampers and transporters, have no basis in known physics. The principles behind them are not even conceivable. However, without inertial dampers, it would take months to speed up, slow down, and make turns, even when traveling at rather sedate speeds like one-fourth the speed of light in a vacuum (see Chapter 10). Travel from the surface of a planet to orbiting spacecraft with any small-sized craft is already nearly inconceivable, so from a scientific standpoint, beaming up via a transporter is not much worse. From a story standpoint, beaming up is far better. It accelerates the plot.

Shields Up!

There is no known mechanism capable of producing the shields depicted in the *Star Trek* and *Star Wars* movies, yet without them every space battle would be a suicide mission. In fact, even space travel itself might be hampered. If a ship finds itself in an asteroid belt or other space junk yard, no problem: up go the shields and all is well.

Since shields are so critical to the entire space-movie genre, let's assume they do exist. A space battle is raging, and a fighter craft is now attacking a large battle cruiser protected by such a shield. The fighter would likely have a mass of around 15,600 pounds (7,091 kg), similar to an Earthbound F-16 fighter, and be approaching kamikaze-style at, say, one-tenth the speed of light—a sedate velocity for spacecraft. The fighter's kinetic energy would be the equivalent of 763 megatons of TNT, or 7.63 of the largest-sized nuclear bomb ever made. Unquestionably, the fighter would be blown up before it got close enough to launch its weapons, but the debris from the fighter would continue forward and impact the cruiser's shields. Naturally, the shields would stop the debris, but what about the kinetic energy they contain? It would have to be turned into something, and about the only choice is heat—enough to convert the debris into plasma—and a giant electromagnetic pulse (EMP) containing every form of electromagnetic radiation from radio to gamma waves.

In the best case, the plasma and EMP would temporarily render the cruiser's sensors useless. In the worst case, it would wipe them out along with the rest of the cruiser's electronic equipment and crew who would be zapped by the gamma rays in the EMP.

Energy, however, is not the only quantity that must be conserved. Momentum also must be conserved. The result is that even if the shields hold, the cruiser is going to get a mighty jolt. At best, the jolt will knock the cruiser out of position; at worst, it will send shock waves into its hull, tearing up equipment and injuring crew members in the process. In short, even if shields did exist, there would still be major problems to overcome.

After the TV series debuted in 1966, *Star Trek* slowly developed as a cultural phenomenon. The TV debut happened at a time of widespread racial inequality—Martin Luther King had delivered his famous "I Have a Dream" address only three years before. Yet not only were the Enterprise's crewmembers multicultural and multiracial, but if a character was Asian, the actor was Asian. By contrast, the famous TV series *Kung Fu*, from 1972, featured a half-Asian, half-Caucasian main character wandering the Old West in America armed only with well-honed martial art skills. A Caucasian with no previous martial arts background, David Carradine, was chosen for the role over world-renowned martial artist, Bruce Lee, an Asian.

Star Trek used space exploration as a stage for exploring all kinds of human issues, including the effects of technology on people. The series exemplified equality and harmony existing in the midst of diversity. Yes, women did initially wear miniskirts, but they were also portrayed in nontraditional roles. Keep in mind that the second-wave feminist movement had just begun around 1960. From the beginning, *Star Trek* was closer to the more forgiving genre of science fantasy than science fiction.

Quick Comparison: Star Wars versus Star Trek

The two franchises include many different media; for brevity, the following will address only movies. There are significant differences between the first series of *Star Wars* movies and the final series, but they are not evaluated separately.

	Star Wars		Star Trek	
	Score	Comments	Score	Comments
Landing and take off from planets	-3	**Large spacecraft**: Land and take off from planets. **Transport of People from Planets:** Shuttle craft with questionable physics	-2	**Large spacecraft**: built and remain in space. **Transport of People from Planets:** Shuttle craft with questionable physics or beaming with unknown physics.
Deadly gs during turning, stopping, & speeding up	-3	**Inertial dampers:** No recognition of problem.	-2	**Inertial dampers:** Based on flaky unknown physics but at least recognizes the problem.
Need for speed	-2	Some vague references to hyperspace. Travel time often unrealistically depicted.	-1	Warp speed concept based on remote possibility of warping space time continuum.
Space battle model	-3	**WWII naval aircraft carrier** (small craft attack big craft loaded with anti-spacecraft weapons): • Small craft must maneuver at slow speed to work with human reaction times and keep gs survivable.* • How could gunners miss? • Every attack a kamikaze attack	-1	**WWII submarine warfare** (big craft attack big craft with 1 or 2 weapons): • Tactics limited • Distances unrealistically close
Artificial Gravity	-3	**Basis:** Unknown physics. **Consistency**: Changes the direction	-2	**Basis:** Unknown physics. **Consistency:** Generally good.
Personal Weapons	-2	**Various blasters:** Conceivable but unlikely **Light sabers**: Use unknown physics and are illogical as primary battlefield weapons.	-1	**Various phasors/disrupters:** Conceivable but unlikely

Figure 38: Star Wars vs. Star Trek

Sound in space	-2	Yes	-2	Yes
Robots/ Androids	-2	**Cutesy Stuff:** Some. **Misc.:** Droid army communicates verbally yet is centrally controlled. Anakin creating C3PO?	-1	**Cutesy Stuff:** Avoids **Misc.:** Data is unbelievably advanced but then he's built over 300 years in the future.
Totals	-20		-12	

* Perhaps high gs would be survivable if the small craft had inertial dampers, but human communication and reaction times would still seriously limit the speed of fighter type space-craft. At a mere 10 percent C, in the time it takes to say, "Use the Force, Luke," he traveled around 37,000 miles (60,000 m). A pilot would travel 1860 miles in the normal reaction time of 0.1 sec. Make an error of a millisecond in pushing a button to launch a missile and it would miss by 18.6 miles.

Figure 38: Star Wars vs. Star Trek Continued

That's not to say that *Star Trek* was always consistent with its own visionary understanding of the future, at least not from a technical standpoint. For example, in *Star Trek IV*, Kirk and crew hijack a Klingon Bird-of-Prey, time-travel back to the twentieth century, land in San Francisco's Golden Gate Park, capture a pair of humpback whales, and fly off with them (a serious break with the tradition of keeping large ships in space) all without being detected by local residents. Okay, the Bird-of-Prey was supposedly cloaked and, hence, invisible when sitting in the park, but does no one walk their dog? Wouldn't Fido be offended by having a spaceship in the middle of his relief area? Wouldn't he, at least, want to mark it? Still, the movie remembers its survival gear, things like scintillating dialog and interesting characters. The movie prospers because it finds that most valued of treasures: humor with heart. It features, among other things, an outer-space alien, Spock—the epitome of logic and reason—wandering around unnoticed in a city with a reputation for creative illogic and a Russian (Chekov) claiming he's a starship officer when caught stealing nuclear energy aboard an aircraft carrier during the height of the cold war—all in a movie about saving whales.

Star Trek's much used WWII submarine warfare model is conceivable but not overly imaginative. According to the model, large ships launch powerful weapons such as torpedoes against other similar ships. While these weapons frequently jostle the inhabitants—similar to a WWII depth-charge attack—the targets are rarely destroyed on the first shot. Such close-range battles would be little more than toe-to-toe slugfests, with few tactical possibilities, although the movies pretend to have them. About the only available battle tactics would be putting up shields and firing weapons. Once in a while, one of the combatants would be able to use trickery or hide in a plasma cloud, but that's about it. More realistic space battles would likely be fought at great distances in a far more imaginative way (see Chapter 5). Still, although the lack of imaginative battles is a disappointment, it's not enough to trigger a fall from grace.

. . . IN A GALAXY FAR, FAR AWAY . . .

On the other hand, the entire WWII aircraft-carrier battle model used by *Star Wars* is flawed, if not outright ridiculous. But in the original *Star Wars* trilogy, it worked. Why? Because the original was an inside joke designed to poke fun at Hollywood. The first movie came out in 1977, only three years after the Watergate scandal compelled President Richard Nixon to resign in shame, and only four years after United States involvement ended in the divisive Vietnam War. At the time America's self-image, as the land of eternal good guys, lay shattered, oozing self-doubt—an image reflected in movies. Characters, even heroes, had to be mixtures of good and bad. Seemingly, nothing could ever again be portrayed in simple terms. *Star Wars* landed on the era's cynical pop culture like an artillery shell. The movie presented everything in the purest of black-or-white, good-or-evil terms. Arguably, the greatest movie villain ever created—Darth Vader—for example, was totally black and totally evil (at least in the first movie). Although Han Solo may have seemed to be a mix, on closer examination he was merely a pure-hearted hero with attention deficit disorder. He might be distracted by personal interests, but given a chance to focus he'd risk everything for the cause.

The original *Star Wars* trilogy was deliberately modeled after obsolete 1930s movie theater serials and used a WWII battle model from the most heroic moment in U.S. history along with light-saber-wielding knights as high-tech updates from classics such as *Seven Samurai* (1954) [NR] and Errol Flynn swashbucklers. These elements were extensions of both the trilogy's positive tone and tongue-in-cheek humor. On that basis alone, movies from the original *Star Wars* trilogy deserve ISMP forgiveness. While they might look like science fiction, in reality they are a mix of parody and fantasy—a humorous yet heroic and altogether ISMP-forgivable mythology.

THE MOTHER OF ALL ISMP LAND BATTLES

Unfortunately, the second trilogy did not follow in the first's footsteps. *Episode I* kicked off the downward decline with a clear case of amnesia. The movie forgot the source of its forgiveness: its roots with the original trilogy. For openers, it offered an unnecessary biological explanation for how the Jedi tap into the Force: "midichlorians," a type of interstellar microbe. The little guys grant access to the Force after planting themselves in one's cells—the more the better. Does this mean swilling a microbe-laced cocktail could make a person stronger in the Force? What about injections? Is the microbe airborne or sexually transmitted, and if so, why are Jedi required to be celibate? Apparently midichlorians fathered Anakin Skywalker. Did they also father the Force, or did the Force father the midichlorians? Was it all some happy cosmic coincidence? By explaining how the Force works, *Episode I* raised more questions than a child does in the fourth year of life. It moved the Force into the glare of scientific and logical analysis and, in the process, evicted the film from the forgiving genre of mythical fantasy.

Having badly weakened its case for forgiveness, *Episode I* proceeded to take one of the goofiest characters ever created, a flop-eared Gungan called Jar Jar Binks, give him a major role, and then build the ISMP classic of all land battles around him and his species. The battle pits the bumbling Gungans against heavily armed, high-tech droids. The Gungans have a sophisticated force-field technology capable of shielding their army on the battlefield, yet they ride around on beasts of burden. They have explosive devices that look like giant blue marbles but have to launch them with ancient-looking catapults. Do they put their knowledge of explosives to work and modify them into propellants for rockets and firearms? No, they use spears. Do they rely on stealth, harassment, or guerrilla warfare—tactics that, at times, have actually worked against technologically superior forces? No, they face off head-to-head with the droids on open ground—a tactic that's usually disastrous when used against technologically superior forces.

When droid tanks fire, their shots bounce off the Gungan force field. Yet droids can walk through the shield effortlessly. Apologists explain that the shields are somehow tuned to block high-energy blasts but allow everything else to pass. Okay, then why didn't a few kamikaze droids loaded with explosives walk through and blow up the Gungans? Why didn't the tanks drive up, poke their barrels through the shields, and blast the Gungans? There are dozens of ways the droids could have improved their battle tactics but didn't.

Yet, even with droid bungling, their superior technology eventually proves invincible: the Gungans face annihilation—then they win. And how does this miracle occur? When the droids' mother ship is destroyed, the droids shut down. Keep in mind that the droids use audible language over their radios to relay information and acknowledge commands. Aside from the illogic of quitting when winning, there was no one in the mother ship capable of giving the command to shut down, so why did they? Apologists answer that given the capability for independent action, droids might have rebelled against their leaders. Yet this doesn't seem to have been a problem for Hitler, Napoleon, Genghis Khan, or most other leaders of major-sized military forces.

Episode I clearly slipped from the state of grace established by the *Star Wars* franchise. On the other hand, with Gungan battles and characters like the flop-eared Jar Jar Binks, the movie looks too much like a Bugs Bunny cartoon to ever be taken seriously.

THE MOTHER-OF-ALL ISMP SPACE BATTLES

Episode II featured the usual assemblage of impossible gizmos, including small-sized craft capable of flying through the atmosphere, landing on their footprint, and making interstellar flights in less time than it takes to drive across Texas. When Senator Padmé Amidala (Natalie Portman) needed to go into hiding to avoid being assassinated, who does she entrust with her senatorial duty of defending the known universe from chaos? Jar Jar Binks (who, unfortunately, was not a candidate for assassination). Fortunately, we are otherwise spared from having to endure him. When Obi Wan Kenobi (Ewan McGregor) wishes to travel to a mysterious distant planet and can't find a record of it in the archives, he consults not the venerable Yoda, but a class of younglings (future Jedi). They give him their profound insight: "someone deleted it from the archives." Gosh, do you think so? And then there is the movie's theme: love is blind. Anakin Skywalker (Hayden Christensen) gets the girl, while incessantly whining and throwing tantrums. (What does Padmé see in this guy?) *Episode II* offered nothing unique with respect to ISMP, but then it offered nothing that made it forgivable. The movie's worst break with its traditions was the inclusion of an unappealing main character—Anakin Skywalker, a mixture of good and evil.

Episode III resumed the downward spiral in its opening scene with what is, arguably, movie history's most ridiculous space battle. The evil General Grievous (voiced by Matthew Wood), along with a major armada of space craft, has somehow slipped into town and kidnapped the head of the Senate, Supreme Chancellor Palpatine (Ian McDiarmid). As the evil general, along with his armada, is escaping into the blackness of outer space high above the capital, Obi-Wan Kenobi and Anakin Skywalker, along with their own armada, race to the rescue.

There's, of course, the trademark WWII aircraft carrier battle stuff, but it's embellished. Obi-Wan, Anakin, and company must fly through exploding flak that leaves little black clouds of smoke. Anywhere near the edge of outer space, these smoke particles would have an outward velocity and essentially no air resistance to slow them down since there's virtually no air. Under such conditions the smoke would almost instantly dissipate. Enemy forces counter attack with everything from vulture-like droids, to droids that fasten onto Obi-Wan's spacecraft and attempt to drill holes in it. Here's a thought: since bullets are cheap and droids are expensive, why not shoot a whole mess of bullets and have them drill the spacecraft? Large spacecraft using eighteenth century sailing ship tactics, deliver close range broadsides with twentieth-century-like cannons, ejecting empty shells out the back as they recoil. Who knows what they shoot, but whatever it is certainly explodes when it hits. Whether or not these were conventional cannons, blowing up an enemy ship at close range would likely be suicidal. The damaged ship's entire fuel supply—an amount designed to provide the humongous energy needs of interstellar travel—could detonate.

During the battle Anakin Skywalker (Hayden Christensen) and Obi-Wan Kenobi (Ewan McGregor) board a huge enemy spacecraft in order to rescue Supreme Chancellor Palpatine. First, the two Jedi fly their fighters through an open entryway and crash-land in a hanger room—without depressurizing the larger spacecraft! After much lightsaber slashing and yada-yada, they rescue the politician just in time for the seriously damaged ship to upend and fall straight toward the planet it had previously been moving away from in an outward, spiraling orbit. (What happened to its orbital velocity?) This dive sends everyone aboard, including the two Jedi, their trusty droid, and the slimy politician they've just rescued, sliding toward the falling end of the ship. Okay, maybe "a long time ago in a galaxy far away" they understood gravity well enough to pump it like central heating fluid through the floors of spacecraft. But if they did, why would the artificial gravity change direction with respect to the floor when the ship fell toward a planet? It seems like the artificial gravity's direction would remain perpendicular to the floor regardless of the ship's position.

Keep in mind that the planet's gravity force never changed direction and cannot be "felt" by an observer on the ship, regardless of whether the ship is falling straight down or orbiting. Both are a form of free fall, which feels like zero gravity. To make things more complex, the spacecraft was attempting to escape into the vastness of outer space. To do this, it would have needed to accelerate until it exceeded escape velocity. The movie depicted the spacecraft as moving in a horizontal direction relative to the ground. As the spacecraft accelerated, its orbit would have tended to spiral outward. People onboard would have felt as though a force was pushing them toward the back of the ship in the opposite direction of the acceleration. An artificial gravity system would not only have needed to compensate for the feeling of zero gravity, but also for the effects of forward acceleration. That's some system! If the ship were damaged so that it started spiraling downward toward the planet, and the artificial gravity was disrupted, the occupants would have floated as though in zero gravity conditions.

Luckily, General Grievous gets the ship back under control, but following another round of yada-yada and light saber slashing, along with taking additional hits on his ship, he decides to abandon it. In the process he jettisons all the escape pods, leaving Palpatine, R2D2, and the two Jedi stranded as the ship once again takes a nose dive.

On descent and reentry the ship—now piloted by Anakin Skywalker—glows red amid a superheated cloud of plasma, breaks in half, and catches fire—yet lands at the nearest spaceport where the plucky rescue team and freed hostage depart uninjured. (If only NASA guys watched movies, just think what they could do.) Can the miraculous descent be explained away by the ship's shields? Not likely; they were at least partly disabled when the Jedi boarded. Can such a wonder ever be explained? Why, yes, it must be the midichlorians!

Okay, the *Star Wars* apologists say that the enemy's armada was not really in orbit but in the extreme upper atmosphere— accounting for the contrast of a black sky with daylight conditions directly below—traveling at suborbital speed. However, the actual kidnapping would have been done by a small group of covert operators. They would have killed Palpatine's body guards, and most likely have been immediately detected. With luck and split-second timing, they could have made their way to a nearby space craft (no doubt a Cosmic Toyota) and blasted off, but not before attracting a swarm of police pursuers. Fearing they'd hit Palpatine, the police would have held their fire, giving the abductors time to travel some distance from the surface. At that moment the previously undetected hostile armada would have dropped out of hyper space (assuming there was such an option) into orbit, and promptly zapped all the pursuers, enabling the abduction to succeed. Needless to say, the armada would be in serious jeopardy and not want to slow down or linger any longer than absolutely necessary before making its way back into space. The idea that an entire armada would land on the surface to kidnap a single politician is preposterous.

Even if the armada temporarily dropped below orbital velocity in the extreme upper atmosphere, staying in a horizontal position without falling toward the surface would require aerodynamic lift or downward thrusters. The thrusters don't seem to be there, leaving aerodynamic lift as the only possibility, but at the edge of the atmosphere there's almost no air. A large spacecraft would have to be going extremely fast and be highly aerodynamic to have any lift at all. If the craft intended to escape into outer space, as mentioned earlier, it would have to accelerate to a speed higher than that required for a circular orbit in order to spiral outward in an ever increasing orbit.

Keep in mind that the apparent lack of gravity for orbiting objects is not caused by being outside the atmosphere, but by having the object moving at the correct speed, in the correct direction, for the given distance above the surface. In theory a spacecraft could be in orbit a centimeter away from the surface, if the planet were perfectly spherical and had no surface imperfections such as mountains. Even if the planet had an atmosphere, a spacecraft with powerful thrusters could overcome air resistance and give its inhabitants the feeling of zero gs just by traveling at the correct orbital velocity in a horizontal direction.

People in an outward spiraling spacecraft would feel as though the weight force were directed backwards, in the opposite direction of the craft's horizontal velocity. With a very slow outward spiral, the apparent weight force would be mild or even imperceptible. The sensation of weight would otherwise not be present.

In the extreme upper atmosphere, opening an entryway large enough to admit fighter craft would not only partly depressurize the larger spacecraft, but likely destabilize it. Air rushing out would act like a thruster which could roll the craft upside down or turn it sideways. If the craft were moving at high velocity in the upper atmosphere, the opening itself would cause a horrendous change in aerodynamic properties, possibly enough to send the craft out of control.

When the craft fell it would still not have fallen straight down, thanks to its high horizontal velocity. Assuming that the craft was free falling, and not wildly spinning or tumbling, its occupants would still have felt weightless during the fall. However, the ships thrusters were firing during the fall, which could have caused the equivalent of a power dive, accelerating the ship downward at a faster rate than the acceleration of gravity. Under such circumstances, people and objects in the ship would have appeared to "fall" upwards toward the tail of the ship. When General Grievous pulled out of the dive, the craft's occupants would have been subjected to accelerations greater than one g. Anything not tied down would have likely slammed into walls.

THE EVIL DATA AND BAD DUNE BUGGY

It would be nice to say that *Star Trek* has retained its originality and remembered that it belonged not just in an action genre, but the last movie installment, *Nemesis* [RP] (2003), gives pause. The movie forgot what an original—let alone, good—script was, and offered a rehash of older *Star Trek* plots—in particular, *The Wrath of Khan* (1982). This time Khan was replaced by a megalomaniac Captain Picard clone called Shinzon (Tom Hardy), who, aside from his shaved head, bore no resemblance to Picard. At best, Shinzon resembled a younger Picard's evil twin. Shinzon takes over Romulus and decides—what else—to annihilate humanity. (If he sounds strangely like a machine, it's because he has the charm of a wood chipper.)

On the way to Romulus, Data finds a missing older brother called B4 (he was made before Data), and completely forgets what happened previously when he found his last missing brother Lore—his evil twin. So what does Data do? Why, of course, he downloads all his memory banks into B4, including exhaustive details of how the Enterprise operates—no risk there. Surprise, surprise, B4 turns out to be a quisling.

Nemesis doesn't feel like a *Star Trek* installment—superstoic Klingon-tough-guy Warf, at a wedding celebration, whining about the side effects of Romulan ale? Give us a break. The movie clearly forgot that the franchise is not about hangover clichés, fight scenes, and other mindless crowd pleasers, but about the human condition and the future impact of technology on the mental and material aspects of our lives. *Nemesis* serves up mostly nonstop action, with everything from hand-to-hand combat, to a lengthy spacecraft battle. The movie even includes a highly contrived "car chase scene." When the *Enterprise* detects B4's signals emanating from a distant planet, Picard thumbs his nose at regulations and drives around the planet's surface in a newly created dune buggy, searching for the signal's source. And why does he insist on personally driving a dune buggy instead of sending others in a shuttle? Is he breaking rules to save his crew or to successfully complete his critical mission? Is he breaking them in service to a higher cause or morality? No, no, and no; he's breaking them because it's more fun. Naturally, bad guys in similar dune buggies give chase, along with copious quantities of poorly aimed blaster fire and wrecks (bad guys only). The away team survives by driving off a cliff, flying through the air, and miraculously landing in the back of their shuttle craft, which has been maneuvered into just the right position by Data using a handheld remote.

THE MOST FORGIVABLE

Both *Star Trek* and *Star Wars* have descended into the depths of ISMP and, at times, forgotten their breathing gear. The *Star Trek* franchise with ten movies and five TV series, compared to six *Star Wars* movies, is the more vulnerable of the two franchise based on size alone. Indeed, *Star Trek* has had more problems with inconsistent script quality than *Star Wars*, but this is not necessarily a compliment. Following the consistently good quality of the first three *Star Wars* movies, the last three have been consistently disappointing.

Overall, the *Star Trek* universe and story lines have been more diverse, and its exploration of future scientific innovations more thought provoking and detailed than those of the *Star Wars* franchise. In their own way, both franchises are gemstones, but *Star Trek* has the edge on deserving forgiveness for ISMP slipups, although its edge is razor thin.

CHAPTER 20

ALL-TIME STUPID MOVIE PHYSICS CLASSICS:
"They Said the Physics Were Impossible . . ."

THE PINNACLE OF BADNESS

It's easy to make a movie with bad physics, but to reach the pinnacle—beyond the merely insultingly stupid—of classically stupid requires artistry. Though many aspire, few achieve. Clearly discernible technical blunders, such as disobeying the first law of thermodynamics, are critical to this category, as is the generous use of bad physics' clichés. Creative physics badness is even better, but the movie must also possess illogical problem solving, inconsistency, and out-of-character or stereotyped scientists, engineers, and other sorts of techies. A few pithy but ridiculous lines of dialogue help immensely, but in the end nothing makes physics flaws shine brighter than a movie which takes itself seriously and is monotonous, clichéd, or lifeless from an artistic standpoint. Mere illogic or exaggeration is insufficient; a true classic must be scientifically unfixable and artistically bland (with the exception of spectacular special effects—these help).

TERRESTRIAL WANNABES

The *Matrix* flirted with becoming an insultingly stupid movie physics (ISMP) classic, but failed. Its explanation of why the machines keep humans as a power source (see Chapter 3) shredded the first and second laws of thermodynamics, and ranks as the ISMP classic for first and second law blunders! The movie followed with mindless nonstop action scenes, but alas, it failed to reach overall ISMP classic status. The first half of the movie was an artistic and science fiction masterpiece, which subsequently spoiled all hope for ISMP classic status.

Matrix sequels fared better, especially the third which served up the ISMP classic of indoor battle scenes: the loading dock battle described in Chapter 5. The unprotected warriors strapped to gigantic robotic devices (called APUs) would have slaughtered themselves with ricochets and falling debris by blasting an impossibly high number of automatic cannon shells toward the concrete roof over their heads. Survivors would have succumbed due to inhaling the smoke from cannon blasts and collateral fires set by exploding shells. Alas, even this nonsense was not enough to grant the movies ISMP classic status.

The pithy line, "they said the physics were impossible . . ." (certainly an ISMP classic), which was followed by defying the conservation of momentum, not to mention the first and second laws of thermodynamics, immediately identified *Eraser* [RP] as a contender. But honestly, even given its physics flaws, the movie has Arnold Schwarzenegger, and who can dislike the big guy. There's a reason he was elected governor and it probably wasn't just his super-duper plan for saving California. The movie was a viable action adventure and its flaws, though grievous, could have been fixed with little (if any) negative impact on the story. The movie failed to achieve the elevated ISMP classic status because its main premise was just not absurd enough.

The Day After Tomorrow's trailer alone created significant Internet buzz. Could this be the one, the ISMP colossus? Certainly, early buzz is a good sign. The special effects budget—a major component of any ISMP contender—was huge. The giant wave crashing against the Statue of Liberty, the LA tornados, and the feeble attempts at dialogue all contributed to its candidacy. It had the enchanted tent. When the movie's hero and a group of friends trudge through a once-in-the-history-of-Earth blizzard to rescue the hero's son, they find solace in the enchanted tent. The wind may howl and the snow may drift, but all is well in the enchanted tent. One can't even see one's breath. This movie looked like a contender. But once again, it was mostly mere exaggeration—nothing especially creative.

DOWN-TO-EARTH OUTER SPACE TRAVEL WANNABES

One might think that near-Earth space travel would offer few opportunities for ISMP classic status. After all, humanity has traveled to the Moon, built space stations, and flown the Space Shuttle. Such ventures have become almost routine. Still, it seems that ISMP opportunities abound even in this almost down-to-Earth environment.

Armageddon certainly rose to the ISMP challenge. It contained one physics flaw after another, mixed with improbability and illogic. It only took a small piece of loose insulation to fatally damage the Challenger, and subsequently over two years to get ready for the next shuttle launch. So how could we possibly train an inexperienced team, have a duel space shuttle launch, land a shuttle on an asteroid, save humanity, and safely return all in eighteen days? The whole idea that humanity can sit and do nothing to prepare for an asteroid strike disaster, then go out and save itself with a puny nuclear bomb blast against a Texas-sized asteroid—all in a matter of weeks—is worse than farcical, it's dangerous. The movie positively overflows with ISMP and improbability. Yet, many of the flaws were simply gross exaggeration. A good number were fixable, and besides, the characters were (unfortunately) likable.

Red Planet (RP) and *Mission to Mars* (MM) ventured a little further into space and contained numerous examples of ISMP. Both have spacecraft with poorly conceived artificial gravity (AG) scenes. MM's AG is merely bad, whereas RP's AG is downright ugly (see Chapter 15). MM offers up an excellent ISMP scene in which a group of astronauts floating in orbit around Mars attempt to rescue Woody Blake (Tim Robbins), who has drifted off, and is at the edge of a rapidly decaying orbit and certain death. To reach him, his would-be rescuers have to continually use their thrusters and are flashed many low fuel warnings by their wrist-mounted computers. But wait, isn't this outer space? What happened to Newton's first law? Once in motion they would have no further need to use thrusters until it was time to stop. RP offered its impossibly flammable alien life forms and homicidal robot, but neither movie had truly groundbreaking levels of ISMP.

Independence Day brings interstellar space travel down to Earth as aliens arrive and attempt a hostile takeover. Even if the movie's premise were accepted—that gargantuan spacecraft could travel to Earth as depicted—when the spaceships were shot down, the resulting explosions as they fell to Earth would destroy the planet's environment (see Chapter 11). There'd be no happy ending. But although the movie delivers some scientific whoppers, it lacks the nonstopbadness that the ISMP all-time classic should possess.

War of the Worlds also showed promise with its alien invasion. Making a gigantic mechanical vehicle—filled to capacity with nasty aliens and high-powered weapons—balance, let alone walk around on the spindliest of legs, would be some trick. Yet, not only did these massive mechanical monsters move, they did so after being stored under a few feet of soil for eons. (Wow, that's some set of batteries!) What's more, they were stored undetected in locations that ended up in the middle of major cities. Did no one dig holes? Did no one have a metal detector? Were there no curious nerds around?

After all that planning and preparation for invading and inhabiting an entire world, these super-intelligent aliens apparently neglected to do an environmental study. What were they thinking? Did they really believe they would be safe? Did they misinterpret all the TV ads for cold remedies and antibacterial soap—intercepted by their version of SETI?

Furthermore, when the aliens do show up in their tripods and start killing everyone, why would a human such as Tom Cruise think he would be safe by fleeing to Boston? Wouldn't heinous aliens bent on turning humanity into fertilizer start with the most dangerous group of all—MIT students? These people are not just nerds, they're uber-nerds. Give them free time and a large-domed building, and for entertainment they do things like put police cars and Wright Brothers' airplanes atop it, or morph it into various new forms, such as a giant pumpkin or an oversized robot. Imagine what they'd do with a bunch of spindly-legged tripods! But disrupting the power for their cool toys (computers), closing down their calculus classes, and threatening the continued existence of humanity is enough to make them mad. Do the invaders recognize this danger and respond by putting the Cambridge and Boston area at the top of their destruction list? No! These aliens are dumber than Tom Cruise's movie character. They deserve to die.

Alas, though the movie offers plenty of ISMP, there are moderating influences. The tripod engineering problems could conceivably be solved by a super-advanced species. Such intelligent critters could possibly even screw up one or two important details when planning an invasion; certainly, such things have happened on Earth. After all, humans—upon sinking the unsinkable Titanic—discovered, to their chagrin, that not only had they failed to plan for enough lifeboats, but had failed to fill even the ones available. In total, *War of the Worlds* just doesn't measure up. Many of its flaws are simple illogic, some are forgivable, and none are creatively spectacular.

AND THE WINNER IS . . .

It seems the choice for the ISMP classic would be tough amid such competition and yet it isn't. There is one that stands at the top of the heap, that is the pick of the pack, and that leads the rest. As mentioned in Chapter 1, its premise is absurd, its science unfixable, and its artistry uninspiring—its name: *The Core*.

The movie explains an impending world horror during a

meeting of top military brass, in which geophysicist Dr. Josh Keyes (Aaron Eckhart) makes a presentation. He tells the military that the core's rotation is messed up and about to cause Earth's magnetic field—all that stands between us and microwave radiation from the sun—to disappear. And why should this cause immediate panic and distress? Keyes, always one to gauge his audience, keeps the explanation moronic: he lights the aerosol from a can of hair spray and torches a peach representing Earth.

A flaming peach? There's a way to impress. Never mind that the audience—leaders of an organization whose technology level and research budget makes NASA look like a club for hobbyists—might actually prefer a few scientific details. Never mind that Earth's magnetic field does nothing to reduce solar microwave radiation arriving at the Earth's surface. Microwaves are a minor part of the sun's total radiation that hit us, magnetic field or not. Never mind that the planet has never been incinerated when it has previously experienced temporary magnetic field loss during the magnetic pole flip-flops that occur about every hundred thousand years. Never mind all the illogic and ISMP, a burnt peach just plain stinks.

Okay, the Sun does periodically emit dangerous quantities of plasma during solar flares, which are made up mostly of protons and electrons with high amounts of kinetic energy. These particles can, indeed, be deflected by magnetic fields. However, even without a magnetic field the atmosphere would still shield us from most of this radiation. Over hundreds of millions of years the lack of a magnetic field could be disastrous. Mars, for example, lacks a magnetic field, which is thought to be the reason it also lacks an atmosphere. Mars' unprotected atmosphere was apparently slowly blown away by solar winds.

The movie continues with other insightful glimpses of scientific and technically trained professionals. Ask any group of high school students to name a recent Nobel Prize-winning physicist and you'll most likely get a blank stare. When it comes to the pop culture scene, world class scientists are a no-show. So, Dr. Zimsky (Stanley Tucci) is, of course, depicted as having a massive ego, star appeal, and the look of a 1930s movie star as he signs autographs and is hounded by a bimbo. There's the genius inventor who lives alone on a vast facility in the middle of the desert and invents technological miracles, all without assistants or outside funding. There are the astronauts who engage in juvenile arguments about who should land the Space Shuttle—an almost entirely automated task. There's the likable philosophical French scientist, with that oh-so-engaging accent—who is also a nuclear weapons expert—who provides just the right amount of sadness when he dies. And, of course, there's the quintessential computer geek, Rat (D. J. Qualls) who can not only hack into a supersecret government facility, but can take over the U.S. power grid, all in a matter of minutes via the Internet. How does this plucky group of technical people solve the movie's dire nonproblem? Why, of course, by traveling in a manned craft that bores into the planet, and restarts Earth's core with a nuclear firecracker (see Chapter 11). An unmanned craft could never suffice.

The Core is a wonder. It would take an entire book to cover all its ISMP details. Will it ever be surpassed? Who can say? But certainly, at least for the foreseeable future, it deserves special status as the ISMP classic. It's so bad, it's good.

NOTES

1. "The Core, A Filmbug Special,"
   http://www.filmbug.com/specials/the core, 3/28/2003.
2. Feynman, Richard, *Six Easy Pieces*, Perseus Books Group, 1996), 69.
3. Frank L. Lambert, Professor Emeritus, Occidental College, "The Second Law of Thermodynamics," http: http://www.secondlaw.com.
4. Ohba, Mitsuru, and Benson, John. "Introduction: About the A-Bomb," http://www.csi.ad.jp/ABOMB/data.html, © 1998 by A-Bomb WWW project.
5. "To the Moon," NOVA, http:www.pbs.org/wgbh/nova/transcripts/2610tothemoon.html, original PBS airdate: July 13, 1999.
6. "The Numbers—U.S. Movie Market Summary for 2006," http://www.the-numbers.com/market/2006.php, 1997–2007 Nash Information Services, LLC.
7. Buell, Harold L, *Dauntless Hell Drivers* (New York: Orion Books, 1991) 37.
8. Jennings, Ed, "Crosley's Secret War Effort: The Proximity Fuze,"
   http://www.navweaps.com/index_tech/tech-075.htm, updated February 1, 2001.

9. "Could a seat belt have saved Diana?" http://www.cnn.com/WORLD/9709/05/crash.analysis/, CNN, original air date: 9/5/1997.

10. "The Freefall Research Page," http://www.greenharbor.com/fffolder/ffallers.html, © 2001–2003, Green Harbor Publications.

11. "How to jump into water from a height," http://www.everything2.com/index.pl?node=How%20to%20jump%20into%20water%20from%20a%20height, updated 11/14/2002.

12. "Could a seat belt have saved Diana?" http://www.cnn.com/WORLD/9709/05/crash.analysis/, CNN, original air date: 9/5/1997; and "Acceleration that Would Kill a Human," The Physics Fact Book, edited by Glenn Elert, http://hypertextbook.com/facts/2004/YuriyRafailov.shtml.

13. Ohba, Mitsuru, and Benson, John, "Introduction: About the A-Bomb," http://www.csi.ad.jp/ABOMB/data.html, © 1995, 1996, 1997, 1998 by A-Bomb WWW project.

14. Levoy, Jill, and Vanessa Hua, "Fighting Bullets on New Year's Eve," Los Angeles Times, 12/30/1999.

15. Yee, Danny, "Average Weight of a Human Head," http://danny.oz.au/anthropology /notes/human-head-weight.html, updated 7/8/07.

16. Zavada, Roland J, "Dissecting the Zapruder Bell & Howell 8mm Movie Camera," http://www.jfk-info.com/zavada1.htm, updated 10/24/98.

17. "Video of Richard Trott shooting a melon," http://mcadams.posc.mu.edu/melon-sh.mpg, accessed 7/8/07; McAdams, John. "Dealey Plaza," http://mcadams.posc.mu.edu/dealey.htm, accessed 7/8/07; Penn & Teller, "Jet Effect," http://video.google.com/videoplay?docid=745248745546892501, accessed 7/8/07.

18. Meyer, Dale K, "Secrets of a Homicide," http://www.jfk-files.com/jfk/html/intro.htmhttp://www.jfkfiles.com/jfk/html/intro.htm, accessed 7/8/07; Jennings, Peter. ABC News, *The Kennedy Assassination—Beyond Conspiracy*, DVD, 2004.

19. Penn & Teller, "Jet Effect," http://video.google.com/videoplay?docid=745248745546892501, accessed 7/8/07.

20. McAdams, John, "The Assassination Goes Hollywood," http://mcadams.posc.mu.edu/jfkmovie.htm, accessed 7/8/07; and Reitses, David. "The JFK 100—One Hundred Errors of Fact and Judgment in Oliver Stone's JFK," http://www.jfk-online.com/jfk100menu.html, accessed 7/8/07.

21. Posner, Gerald, *Case Closed, Lee Harvey Oswald and the Assassination of JFK*, ([AU: Please provide place of publication]:Anchor Books, 2003), 273–284.

22. The apparent zero gravity occurs because the orbiting object is in freefall. If an object has a horizontal velocity, it will fall in a parabolic path and hit the ground, but if the horizontal velocity is high enough and there's a significant gravity force to act as a centripetal force, the object will fall in a stable orbit and never hit the ground (assuming no air resistance).

23. "California Tornadoes 1880–2000," http://www.tornado-project.com/alltorns/catorn.htm, the Tornado Project, accessed 7/8/07.

24. "The Staten Island Web," http://www.si-web.com/Statue.html, accessed 7/8/07.

25. Ohba, Mitsuru, and John Benson. "A-Bomb WWW Museum," http://www.csi.ed.jp/ABOMB/data.html, © 1995, 1996, 1997, 1998 by A-Bomb WWW Project.

26. "If the Polar Ice Caps Melted, How Much Would the Oceans Rise?" How Stuff Works Web site, http://science.howstuffworks.com/question 473.htm, (c) 1998–2007 HowStuffWorks, Inc.; "Volume of Earth's Polar Ice Caps," The Physics Factbook, edited by Glenn Elert, written by his students.

27. http://hypertextbook.com/facts/2000/HannaBerenblit.shtml, 2003; and "Power Density of Solar Radiation," The Physics Factbook, edited by Glenn Elert, written by his students. http://hypertextbook.com/facts/1998/ManicaPiputbundit.shtml, 1998.

28. Gleick, James, *Chaos: Making a New Science*, ([AU: provide place of publication]: Penguin Books, 1988).

29. "What is the Temperature at the Tip of a Lit Cigarette?" Physics and Astronomy Online, http://www.physlink.com/education/askexperts/ae1.cfm, © 2007 PhysLink.com.

30. "Ignition Temperature of Gasoline," The Physics Factbook, edited by Glenn Elert, written by his students, http://hypertextbook.com/facts/2003/ShaniChristopher.shtml, 2003.

31. "Reducing the Health Consequences of Smoking—25 Years of Progress: A Report of the Surgeon General," 1989 Executive Summary, 79.

32. Gustin, Emmanuel, "The Fighters," http://users.skynet.be/Emmanuel.Gustin/fgun/fgun-fi/html, © 1998–1999.

INDEX

ABOUT THE AUTHOR

Tom Rogers is the founder and creator of the wildly popular website Insultingly Stupid Movie Physics. He has a bachelors degree in mechanical engineering from Arizona State University and a Master of Business Administration degree from Clemson University. He worked as an engineer for eighteen years and currently lives in Greenville, South Carolina.